IRATIONAL.ORG'S TRAUM

A

PSYCHOARCHAEOLOGIST'S

DRAMATURGY

V.M

A collaboration between the Post-Media Lab
& Mute Books

Mute

Anti copyright © 2014 Mute

Except for those images which originally appeared elsewhere and are republished here, all content is copyright Mute and the authors. However, Mute encourages the use of its content for purposes that are non-commercial, critical, or disruptive of capitalist property relations. Please make sure you credit the author and Mute as the original publishers.

This legend is devised in the absence of a licence which adequately represents this contributor's and publisher's respective positions on copyright, and to acknowledge but deny the copyrighting performed by default where copyright is waived.

Please email mute@metamute.org with any details of republication

Co-published as a collaboration between Mute and the Post-Media Lab, Leuphana University.
PML Books is a book series accompanying the work of the Post-Media Lab. http://postmedialab.org

Print ISBN: 978-1-906496-98-2
Also available as eBook ISBN: 978-1-906496-78-4

Distribution Please contact mute@metamute.org for trade and distribution enquiries

Acknowledgements

Series Editors Josephine Berry Slater, Anthony Iles, Clemens Apprich and Oliver Lerone Schultz
Book Concept Rachel Baker / Vahida Ramujkic
Layout Vahida Ramujkic / Raquel Perez de Eulate
Design Template Based on a template by Atwork
Cover Image irational.org, *Forest Bureau*, 2013

PML Books

The books in this short series are:

Claire Fontaine, *Human Strike Has Already Begun & Other Writings*, (ISBN 978-1-906496-88-3)

Felix Stalder, *Digital Solidarity*, (ISBN 978-1-906496-92-0)

Clemens Apprich, et. al., *Provocative Alloys: A Post-Media Anthology*, (ISBN 978-1-906496-94-4)

Clemens Apprich, et. al., *Plants, Androids and Operators: A Post-Media Handbook*, (ISBN 978-1-906496-96-8)

Rodrigo Nunes, *The Organisation of the Organisationless: Collective Action After Networks*, (ISBN 978-1-906496-75-3)

The PML Book series is just one of several outlets for the Lab's exploration of post-media strategies and conditions, which includes fellowships, a virtual lab structure, multiple collaborations, events, group readings and other documentation.

For more information see: www.postmedialab.org/publications

MUTE BOOKS

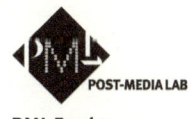
PML Books

The Post-Media Lab is part of the Lüneburg Innovation
Incubator, a major EU project within Leuphana University of
Lüneburg, financed by the European Regional Development
Fund and co-funded by the German federal state of
Lower Saxony.

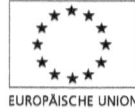 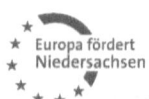 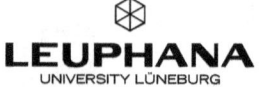

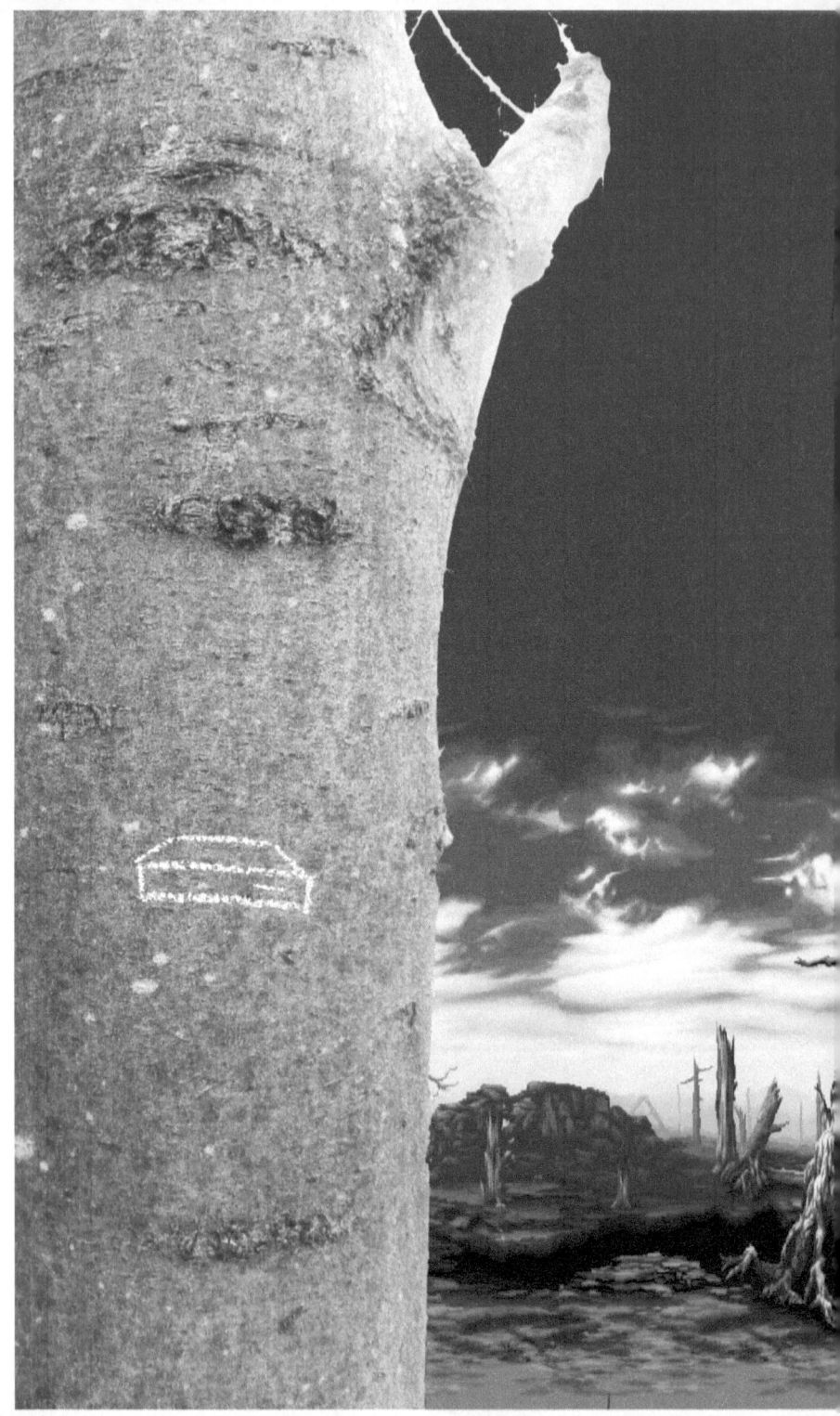

OBJECT) Precious
fictional
mamasties.

MYTHOLOGY

convey laying

ART W.
EV.

CREDIBILITY

collectors
op briars

A Letter To The Publisher

V.M. Psychoarchaeologist
February 2044

Dear Sir/Madam,

Regarding our conversation about late 20th and early 21st century autonomous server communities. I enclose here unfinished notes and data artefacts from my irational.org log book.

I've attempted, unsuccessfully perhaps, to reconstruct this pre-NullNet society. But I feel this mosaic of reproductions is so incomplete as to lead only to further frustration. My attempts have led me to write some dramaturgical scripts based on meetings that appear to have taken place, a technique I often use as part of my research methodology. My brief encounter with the group whilst a student at Leuphana University in Lüneburg has also contributed to this assemblage of dreams, mind games, seizures, allusions and delusions.

The primary excavation method available to me is through sending multiple requests for data artefacts by land mail to 'The Thief' using the Secure Access Protocol form. It takes months for these submissions to generate a response and if any material does arrive it is often unintelligible, detached as it is from its original online context. And it is not at all certain that these are genuinely derived from the original server.

The Thief alleges that he is the last remaining User of irational.org before the internet was nullified, and also claims to be in possession of a copy of the archived disk. There is no way to verify this as I simply do not have the resources to track him down and investigate. All I have is a P.O. Box address. He has appointed himself Guardian of the disk and assumed duties as the official

Data Auditor, so we are largely at his whim in recovering the material in a fragmented and piecemeal way. In addition he has redacted many texts. This implied sensitivity attached to the data suggests that unknown irational.org personae are still vulnerable to some kind of privacy infringement leading to incriminating exposure.

There are so many forms to deal with. It is as if the irational, authoritarian, impersonal language of the form had overtaken all other, and knowledge of a thing or person can only occur through this mechanism.

Of course, post-internet, we are still familiar with the tyranny of the form, the office, the structures and tools of bureaucracy. But in that haute-internet age, when state and corporate exploitation of privacy was about to bring the crypto-libertarians into direct confrontation with the liberal transparency advocates, there was a deep crisis of administration and information management. It was during the Document Dumps and Whistle-blowing Wars of the 2020's when the micro-infrastructure of servers echoed the perfect surveillance infrastructure of the state as it was, engendering a deep crisis of power relations.

With irational.org, the attempt to repel or escape this administrative teleology and its institutional power relations is evident, whilst at the same time internalised to such a degree that its logic consumes and negates their own social organisation. Ultimately they succumb, and there is nowhere left to flee but the ex-urban wastes, and the forest. But, as you will see, there is evidence to suggest that the irational group may also have been planning to migrate their server architecture into the physical properties of a public park. Using the disk archive as a schematic planning tool it appears that an irational Park was being designed as some kind of mirror or echo of the server. Had the group anticipated the internet meltdown? Were they meeting to plan an escape from the world of electronic bureaucracy? Did irational.org disk data become inscribed into the park somehow? There is not enough evidence to know for certain.

The dramatis personae of the server are rendered as functions, rather than attributed with names. I have no way of knowing if the Thief is responsible for this. It is coterminous with his theatricality of anonymity, privacy and paranoia. They have had their names deleted, but their identities and interrelations are interpreted through various emails, texts and transcribed audio found in the archive. The nature of discussion is playful, often ludicrous. Delay, deferral and misdirection is employed with respect to accessing the disk archive. I speculate, but in these evasions they do seem to achieve transcendence from the server bureau, eventually metamorphosising it into either a park or forest, or both.

Emerging from these server remnants is the longing for human companionship, the struggle for art in a cultural environment built on the hierarchical administration of the computer, the need to unify the body with nature. Absurd administration itself becomes the source of the poetry. The quest to access the archived server proved unattainable, but in the process of excavation I became deeply involved with the lives of the users of the irational.org server and their escape from it.

Given their fragility and sensitivity I offer these fragments for you to publish at your own risk.

Yours
V.M.

SECURE ACCESS PROTOCOL

REF: IR001
NAME: SECURE ACCESS PROTOCOL. DATA PRIVACY POLICY AND REQUEST FORM.
TYPE: TYPED DOCUMENTS x 2
DATE: MID/LATE 2013

OBSERVATIONS: Irrational.org archived its server onto a hard disk.
The Secure Access Protocol is an indication of:
  a) security as a high priority during the period of state and corporate surveillance, and
  b) the security of data.
It anticipates the value of the disk as a cultural object.
The Thief has retained the SAP system but in a post-internet world the process must adapt to analogue transfer of materials.

QUESTION: Did Irrational.org predict the Internet crash? Was it planning to leave the server behind and migrate/transform data into a PARK?

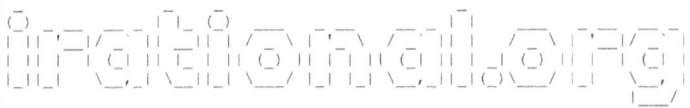

Data Object Access Request Form

This form can be used to request material relating to activities by irational.org as documented on the their server. Although we are not legally obliged to provide this information we'll do our best. All requests must be made in writing, providing a name and address for correspondence and describing the information sought. You may be asked for evidence of your identity. This is to make sure that personal information is not given to the wrong person.
If you would like to request information from **irational.org**, then please complete this form and return to the **Information Compliance Unit** at **xxxxxxxx**, xxxxxxx xxxxxxxxx xxxxxxx xxxxxxxxxxxxxxx 1, xxxxx

YOUR DETAILS

Full Name	
Address	
Tel. No	
Email	

DESCRIPTION OF INFORMATION YOU REQUIRE

Please provide a description of the data you would like to be provided with including relevant details on format (web pages, documents, images, audio, video, scripts or emails) or other (dates, author, project etc.)

Please mark your preferred mode of information release with an X

	View by appointment
x	Printed copy
	A verbal summary
	~~An electronic copy, sent by email~~

NOTE ON REDACTION

All information requested may be subject to redaction under the terms of the Data Protection Act in order to protect third parties. The irational.org **Data Auditor** will analyse the information before release and will redact accordingly.

Secure Access Protocol: Data Protection for the artserver Irational.org and its 2011 disk archive

Irational.org is dedicated to safeguarding and preserving the privacy of its members, associates, friends and all those who have visited our site and communicated electronically with us, in the past, present and future. In 2010 we made a copy of the server.

This Data Protection Policy explains what happens to any personal data that you have provided to us, or that we have collected from you when you interact/ed with Irational.org, its members and projects. We may update this Privacy Policy from time to time. For the purpose of the Data Protection Act 1998 our data controller is: THE DATA AUDITOR.

Information we collect
In interactions with Irational.org, we may have collected and processed the following data about you:

1. Details of your visits to our website and the resources that you access, including, but not limited to, traffic data, location data, weblogs and other communication data.

2. Information regarding your computer whilst you were on our website.

3. Information that you provide by filling in forms on our website, such as when you registered for information or made a purchase.

4. Information provided to us when you communicated with us for any reason e.g via email.

Use of your information
The information that we have collected and stored relating to you is most likely accreted data that has been forgotten or neglected as historical mould, but there is a chance that it will be used:

1. to provide information on irational products and services which we feel may be of interest to you.
2. in a project for exhibition to meet our contractual commitments with commissioners.

Storing Your Personal Data
The transmission of information via the internet is not completely secure (link to GCHQ) and therefore we cannot guarantee the security of data sent to us electronically and transmission of such data is therefore entirely at your own risk. Where we have given you (or where you have chosen) a password so that you can access certain parts of our site, you are responsible for keeping this password confidential.

Disclosing Your Information
Where applicable, we may disclose your personal information to any member of our group, subject to DATA AUDITOR arbitration.

Third party links
You may find links to third party websites on irational.org. These websites should have their own data protection policies which you should check. We do not accept any responsibility or liability for their policies whatsoever as we have no control over them.

Access to information
The Data Protection Act 1998 gives you the right to access the information that we hold about you. Should you wish to receive details that we hold about you please contact us using the Irational.org Data Subject Access Request Form. Should you wish to receive information relating to activities by Irational.org, please contact us using Data Object Access Request Form. We are not legally obliged to provide this information but we'll do our best.

Job description for **root@irational.org / root** on a collective server.

root is a multiple-use name in linux. It is used, and in more radical server environments shared, by server administrators worldwide. A multiple-use name or multiple identity is a name used by many different persons. Other multiple identities include Anonymous, Luther Blissett and Buddha, which is both a proper noun and a condition that may be achieved by anyone. On linux anyone could be **root** just by typing the **root** password. However not everyone can be **root**.

root can be described as a role where an individual user(s) is granted permissions to act in a certain capacity, either by request, choice or circumstance. Member of the Public is another widely recognised role in society, although most people are never elected to be members of the public, only when they die tragically or act heroically are they elevated to such a high status. Most of the time most people are private individuals, even when in public space.

Irational.org currently has three **root** users who manifest as a single super-user who can see and do anything/ everything. When logged in as **root** you technically have authority over the whole machine. You can for example type a single command/make a simple typographic error and destroy the whole system. In this job a thirst for restraint and lack of curiosity are a bonus. As a multiple identity - unlike regular users who are often (although not always) individual persons - **root** will aspire to generic behaviour. For this reason customisation of **root**'s environment is an anathema.

Simultaneously, due to holding ongoing responsibility for the integrity of the whole machine, all users and data, **root** must be poised to defend against external attack as well as breakages, decay and other inhouse failures. A paranoid attitude is justified as you will be actually and specifically traversing a meteor storm of attempted attacks on your server as a daily event. There is no real parallel in normal life. Accordingly **root** must maintain conflicting tendencies for inertia and dynamism and be prepared to accelerate from one state to the other at the touch of a button.

Data Protection

Security is a component of data protection actioned daily. As **root** it is your job to protect both users and the operating environment from a variety of credible threats. The source of these threats will range from malicious hackers to oneself.

For starters, **root** has the technical ability to read all
data in the system. Regular users do not automatically
receive privileges to access data other than their own
unless explicitly granted, and are often unaware or
oblivious of **root** unless they need something done, leaving
root with free rein, a cat amongst pheasants. Having
multiple users with **root** access is one way to limit the
linux operating system's inbuilt bias towards dictatorship,
it is also more convivial in the face of tedious admin or a
disaster.

While **root** inhabits a natural state of despotism, the job
description is probably closer to cleaner or janitor, in
that while you have keys to everyone else's room there are
clear protocols for entry, strictly on
invitation of and under dweller's instructions (although
with implicit permissions in cases of emergency); and with
the imperative to pretend to not observe anything unless
directed. This analogy might be insufficient
though, as cleaners generally have many keys but are not
granted the master key.

In light of contemporary revelations over global
surveillance apparatus run by the NSA and numerous
commercial and international partners, it can be speculated
the NSA is **root** of the internet.

Reliability

Reliability as a concept includes the ability to
work under stress and learn quickly, being loyal and
conscientiousness, and having a diligent and thorough
work style. As **root** it is good practice to have a well-
structured file plan and standard file-naming conventions
for electronic documents, along with an aptitude for
pattern seeking, the ability to parse and filter too much
information, and an exaggerated respect for due process.
The term reliability also refers to compatibility between
root-specific tasks and his/her other primary and secondary
duties, as a normal user and regular account holder on the
server &/or member of the art collective

Independence/Neutrality

It is important that **root** be able to satisfy its data
protection duties whilst also avoiding conflicts of
interest. An internal conflict of interest can be avoided in
theory by appointing an external**root**, for example a **root**
swap with another, similar server. However, appointing an
external**root** is no actual guarantee against conflicts of
interest, particularly in the sphere of networked media art
where everyone is potentially or actually related, making
the condition of impartiality moot.

root@irational.org Jan 2014

BAD LIBRARIAN

BAD LIBRARIAN
Now you are here I'm afraid I must tell you the plain truth – we don't need you as a user.

USER
But I was invited here to research the disk.

BAD LIBRARIAN
There isn't any use for you here as a researcher. The boundaries of our archival content have already been mapped, assessed, analysed, interpreted, recorded. Besides, there's too much sensitive data here. It requires a lot of security.

USER
This is a great surprise to me. I can only hope there's been some misunderstanding.

BAD LIBRARIAN
I'm afraid not, it is as I say.

USER
I haven't made this journey here just to be sent back again, have I?

BAD LIBRARIAN
That isn't for me to decide. But maybe I can explain how this misunderstanding came about.

Secure Access Protocol

USER

Yes, please.

BAD LIBRARIAN

A while ago a proposal for a Secure Access Protocol was issued to enable safe user access to the disk archive. I can search for this proposal and show you. I used to keep everything, but I don't remember the filename, or the folder. You must realise, much work has been done to build this archive and its documentation, this represents only a small part of the total. Much of it has gone missing. The proposal is a document with the word Secure Access Protocol written as blue type in the content-body. Or is it green...?

USER

It all seems a bit chaotic.

BAD LIBRARIAN

It's impossible to keep up with everything. So much remains undealt with. And especially when I become ill, it all gets out of hand.

USER

Couldn't I help you look?

BAD LIBRARIAN

No. I cannot let you look through the archive yourself, that would be going too far.

USER

Don't you have any archive assistants?

BAD LIBRARIAN
Yes, they follow me around and get in the way.

USER
Not much thought has gone into their employment then.

BAD LIBRARIAN
Nothing here is done without thought.

USHER
What about my invitation here as a researcher, as a user of the archive?

BAD LIBRARIAN
That was also given a lot of careful consideration. But incidental circumstances caused confusion. The document will prove it to you.

USER
Where is this document?

BAD LIBRARIAN
I'm not sure, but anyway I can explain what happened without the document. We replied to the proposal for a 'Secure Access Protocol', saying it was impossible. The reply did not reach the original Department. Lets call it Department A. By mistake it went to Department B, but it was incomplete, that is, only the covering note arrived at Dept B. So the official there sent notification that completion was necessary. Many months had passed since the first communication from Dept A, so everyone only had a vague memory about the original proposal for the Secure Access Protocol, so could only

reply with a request for more information about it and why there was a need for one. Naturally they were not satisfied with this reply. A long correspondence began. To the enquiry as to why a Secure Access Protocol was required in the first place we answered that we assumed the idea had originally come from the Head of Security. But as this was a different Dept. than the one originally concerned we had long forgotten. The first official proposal was demanded but this had been lost. Needless to say there followed accusations of plots, injustices and conspiracies. And so, a matter of practicality – the need for a Secure Access Protocol – became a matter of doubt. The Head Of Security then refused to have anything to do with the Secure Access Protocol. But in the meantime the archive administration...

USER
Oh, do you know the Office of Archive Administration?

BAD LIBRARIAN
Oh yes, of course, that is the most important office. They discovered that communication had been sent to Dept A concerning the Secure Access Protocol and had sent a query to the Dept but with no result. A new query was sent to me and I eventually cleared up the whole business. So imagine my dismay at your appearance here now. You are a security risk.

USHER
But here is my invitation!

BAD LIBRARIAN
I'm not authorised to comment on that. But it doesn't

look like an official document. And there is no mention of the archive. So, it seems I cannot find the document right now. But now you know the story, so we don't need the document anymore. It's sure to be found anyway, probably at the Office of Administration, they have all the documents.

USHER
Do you know the Secretary to the Chief of Administration?

BAD LIBRARIAN
I did have dinner with the Chief once. But you cannot expect me to know all the Secretaries.

USER
They passed an email to me indicating that I was invited to access the archive.

BAD LIBRARIAN:
Unfortunately, although I am concerned with official matters I have no telephone to contact the Chief of Administration to confirm this.

USER
Everything appears unclear and insoluble, except that I will be dismissed.

BAD LIBRARIAN:
Who will dismiss you? The lack of clarity in the affair guarantees you the utmost courtesy. Nothing is keeping you here, but you are not being thrown out.

Secure Access Protocol

USER
But what about my long journey here?

BAD LIBRARIAN
I will report it to the Chief of Administration, if a decision is made I will notify you.

USER
I don't want any favours from you, I just want my user rights.

V.M. With acknowledgments to Kafka's 'The Castle'

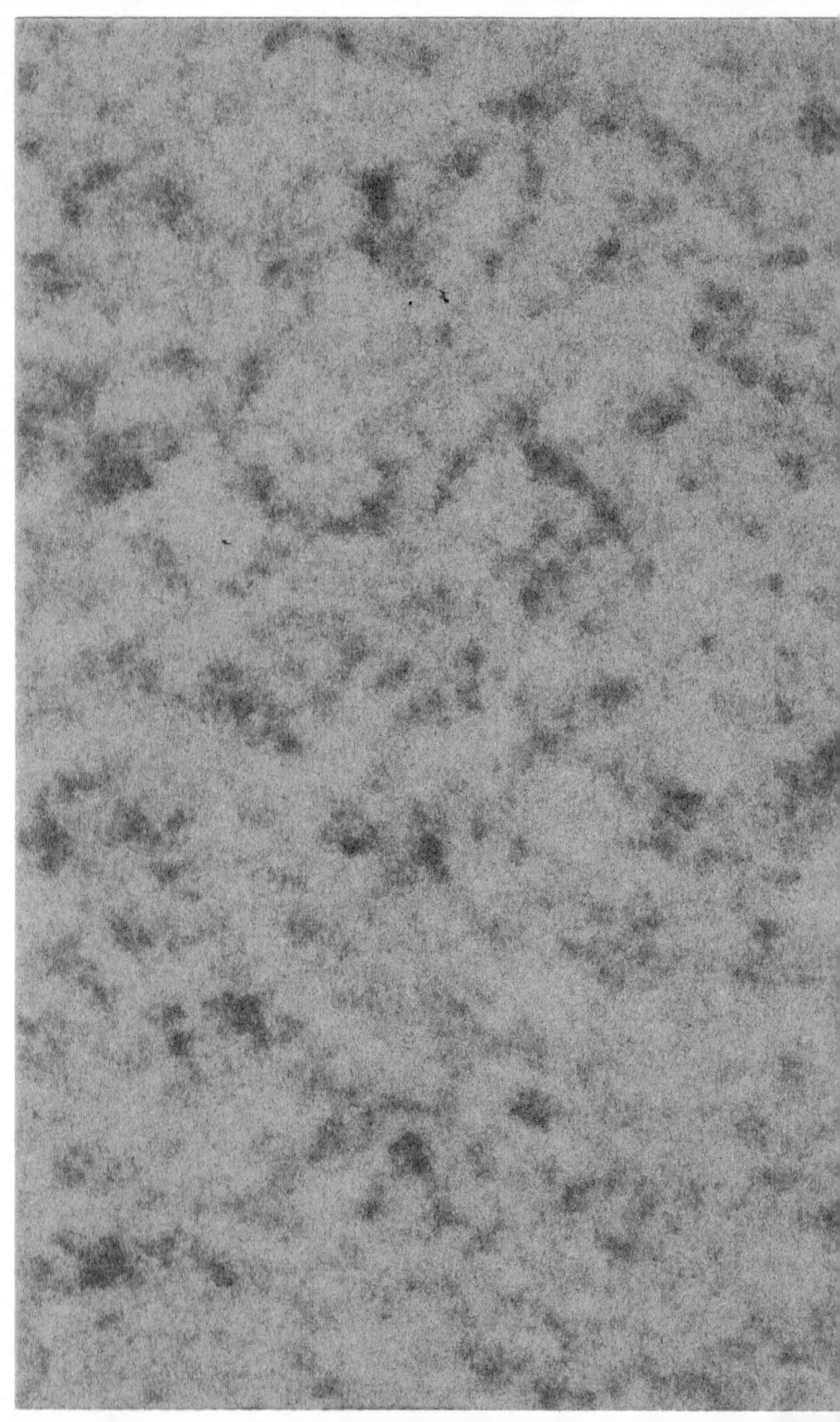

ANTI-ADMINISTRATION

Ref: IR 0004
Name: IRATIONAL.ORG CONSTITUTION
Doc-Type: TRANSCRIPTION OF MEETING
Date: 29.03.2013

Members meet to draft a constitution to satisfy the requirements of a bank account.
The purpose of Irational.org remains unclear, but the idea of The group as a fire around which a group assembles to keep warm as symbolically significant.

Subject: irational.org constitution (fwd)
Date: 29.03.2013

Treasurer: we need to create a constitution to open a bank account in xxxxx. we will use a constitution of a local art group - xxxxxxx kunstverein. what can we keep? what can we delete?
Sysop: remove sections about the board
Treasurer: all members are directors? shareholders? a co-op?
Sysop: none of these if the organisation is not formally constituted/issuing shares. 'members' is more accurate. keep reference to 'treasurer' only
Treasurer: also, the elephant in the room is that i am the founding member, a kind of de facto leader
Sysop: we could create a position of president or CEO, that could be rotated every two years, who would deal with constitutional matters, and inter-corporate/collective relationships
Treasurer: that should be everyone
Sysop: 'should' is based on wish, not reality. simplify expulsion process to consensus
Sysop: do we want to open membership to other corporation/collectives?
Treasurer: probably not
Sysop: the purpose of irational is still entirely unclear to me so this is the first thing that needs supplying. First: The association pursues exclusively and directly intended to ....
Treasurer: ...to be an association to satisfy the basic human need of membership of a group
Cleaning Lady: this sounds a bit strange: what are the basic human needs that an association could satisfy? to be member of some group?
Treasurer: its a basic human need to belong socially.
Cleaning Lady: also, should we mention things like 'tree climbing day' as an established holiday, etc?
Sysop: good idea. would you like to put that in more constitutional language?
Sysop: some unanswered questions: where is irational.org located?

Cleaning Lady: can we say - transnational forest?
Sysop: we could. a bit obscure though. saying THE FOREST is weird enough
Treasurer: i would say both. you could add 'including adjacent waterways'
Gardener: is the membership of the collective limited by NUMBER?
Cleaning Lady: i think not
Treasurer: i would say 8 is a good number of people
Sysop: current server could probably support 200 users, seems like a dangerous limitation. also electricity might not exist as long as irational. the server should not be a collective requirement
Treasurer: we could also do it by total kilogram though
Sysop: 600 KG? clothed?
Treasurer: without shoes is normal for weighing
Secretary: how about 'limited by the server capacity and processing power'.
Treasurer: irational was initially conceptualised around the server, but with the growing availability of public internet services we re-conceptualised ourselves around common work themes i.e contestation of property and representation. we need to be able to regularly change how we define ourselves. i propose our primary conceptualisation as providing the framework for members to satisfy their human need for group membership. the fire in the forest could eventually replace our server
Sysop: on the dissolution of IRATIONAL where do its assets go - to MEMBERS?
Cleaning Lady: how other to deal with common propriety?
Treasurer: or how about the United Nations ?
Sysop: i would be unwilling
Treasurer: what do you suggest ?
Sysop: something more foresty. Unesco?
Treasurer: sounds good
Sysop: so

§ 2 Purpose of the Association. First - The Association is directly intended to ...

Treasurer:
1. TO SATISFY THE BASIC NEED OF BEING A MEMBER OF A GROUP
2. TO CONTEST PROPERTY AND REPRESENTATION

Sysop:
> § 3 Membership.
>
> First - Member of the Association is open to any individual who also is a legal entity of public or private law and a human being.

Cleaning Lady: what is 'legal entity of private law'? Seems irrelevant to me.

Sysop: the appointment of new members must be submitted in writing.

Cleaning Lady: ...a motivation letter?

Sysop: means that the decision to add a new member should be written down, not just spoken. seems reasonable
> Second - The Collective decides on the appointment of new members at its discretion. In rejecting the application, it is not obliged to inform the applicant of the reasons.

Cleaning Lady: I wouldn't put this.
> 3rd Membership is terminated by death, withdrawal from the association, deletion from the membership or exclusion by consensus.
>
> 4th A member may be expelled from the organization, if in a damaging way it has violated the interests of the association. The exclusion is at the request of a member of the organisation decided by consensus, or if in a situation of deadlock by voting. The decision of expulsion is communicated to the member in writing by the General Assembly and shall take effect upon receipt by the member..
>
> 5th Which shall, upon termination of membership is not entitled to a share in the assets of the association.

Cleaning Lady: but, if someone has invested for years in something s/he should be paid indemnification.

Sysop: 'not entitled' means no necessary entitlement. reversing that in the constitution is difficult for a lot of reasons as it has to cover all eventualities. for example if a member was expelled due to theft

– also imagine the organisation's assets are 1 x server & 0 other assets. the departing member claiming their 'share' of the server would be very destructive. goods held in common are not designed to be divided up. also as far i can see it, irational is not for holding assets but running utilities (currently the server, & enough cash to protect its immediate future). any extra assets that come in are shared out in cash as we go. the investment is social not monetary. putting in the constitution the idea you would cash out when you leave i think is a bad idea.

Cleaning Lady: i agree if there is no other assets, apart from the utilities, and there is nothing to be shared or paid out. i am just thinking about it as a model that could fit different situations.

Sysop: not sure we should make it part of the constitution now, when it doesn't match the current or past activity.

§ 4 Membership Fees.

First – Members pay a monthly membership fee.

Second – The amount of the contribution as well as of fee reductions is decided by the Treasurer with consensus of the General Assembly.

Third – Honorary members are exempt from the obligation to contribute.

Treasurer: might be better to call these 'users'.

Sysop: if you prefer. but the word 'user' does imply structure of a server/computer, i think honorary member is broader & good

Cleaning Lady: honorary members should also be decided by consensus.

Sysop: right now people are free to add friends & colleagues as email/server users under their protection just by informing the sysop. do we really want to have everyone approve these first?

Cleaning Lady: ok, but if people are allowing disk space for other peoples' projects, is there any limit of how much space each member could use, how many new sub-domains and users can be made?

Sysop:

Fourth The Treasurer will collect, hold, transfer and disperse the common funds in appropriate currencies, in the first place for the upkeep and continuation of IRATIONAL.ORG property and infrastructure; thereafter for the benefit of the Association's members and the collective purpose.

Cleaning Lady: ...and inform about common funds to the collective quarterly (every three months)

Sysop:

Resolutions including constitutional amendments are passed by consensus. Abstentions in the form of silence are not counted.

Treasurer: I think we should not assume silence is abstention. we should assume silence is acceptance but not necessarily support of a proposal. in my mind, silence is not abstention automatically. in many cultures, silence is active not passive.

Sysop: ok then. Abstentions in the form of silence will be counted. consensus is not everyone saying 'yes', its just not someone saying 'no'.

Treasurer: participation in the form of silence will be counted as acceptance unless the security of irational is in jeopardy. response latency will not be interpreted as silent acceptance

Secretary: so whats the acceptable period of silent time before agreement is assumed and action taken? 5 days? 3 days? 24 hours? 1 hour..?

Treasurer: that will be in constant flux and requires some delicacy of judgement. coding such precision into a fundamental document will probably render it dysfunctional

Cleaning Lady: actually i just thought it is good thing for a collective to have some discussion

Sysop: in the absence of further comments, can we go with this for now?

Treasurer: sure - it can always be amended

REF: IR005
NAME: LAWYER, SPEECH
DATE: 1996/2013

This transcript records a speech delivered by a lawyer at the Institute of Electrical Engineers, London in 1996.

It discusses privacy infringement, and it's publication here, and on the www, is an infringement of privacy.

Secretary was employed there creating a interloper temp network at whistleblowers.

Converted *.ra file – 1996

"The LAWYER Transcript"

CHAIRMAN

Our next speaker is going to give a talk entitled 'Engineering and Law – Compatible or Contradictory?'
Brian Niblet is a barrister and a computer scientist. He was for several years a tenant of the intellectual property chambers in the Temple and has been a professor of computer science at The University Of Wales.
...Brian!

Everyone claps

BRIAN NIBLET

Like the other speakers here today, I'm delighted and honoured to be here.
This is a conference on Licence, Liberty and Privacy. I have been to a number of conferences on privacy and I have to tell you that conferences on privacy are dangerous.
Every privacy conference I have ever been to has infringed privacy.
You see, the first thing that happens is that they gave you a badge – a label.
You must be identified at a privacy conference.
Now I don't wear badges.
But it doesn't help me.
Because I'm always identified as the man who doesn't wear a badge at privacy conferences.

Audience laughs

Then there is the danger to the speakers.
You see, when I was invited to give a talk here they said 'of course your talk will be recorded, it will be transcribed, it will be published'.
That's what happens at Privacy conferences.
Now I have an agreement with the organisers here, a legally binding agreement, a contract indeed, that my talk will not be transcribed.
And I wanted this, because I believe in privacy.
Privacy is freedom.
I want the freedom to talk today at this closed meeting without being transcribed and published - I want that freedom.

The second reason I negotiated a contract with the organisers was for this reason: I wanted to demonstrate to you that the contract, that legally binding agreement, a private treaty between two parties, is one of the effective methods of protecting privacy.

And it's protecting my privacy today, limiting my talk to the audience which is sitting here.

Now, it's a feature of regulation that it does infringe privacy.

I want to tell you today that the most effective method of protecting privacy is the Common Law Of England.

Text on Screen: **REGULATION INFRINGES PRIVACY**

Regulation infringes privacy.
Let me draw the distinction between Common Law and Regulation.
Common Law is the aggregate of commands developed by induction in a process based on reality.
Common Law is anchored in reality - the judges deal with actual facts and actual cases and actual circumstances and from that they abstract a principle which is grounded in reality.
It is a bottom-up process.
Regulation is a top-down process, it never reaches the ground.
It is bureaucrats and legislators.
It is not in touch with the ground.
Regulation infringes privacy quite often.
And the extreme example of that is the first statute in this country said to be concerned with privacy - **The Data Protection Act of 1984.**

The Data Protection Act of 1984 is a regulatory measure. It infringes privacy therefore.

The first thing it does is have a regulatory Officer - The Data Protection Registrar.
Then, it has a register of every data user and computer bureau in the country.
You must be identified. You must wear a label.
In that register you must disclose every purpose for which you are going to use that personal data.
If that is not an infringement of privacy I don't know what is.
And if you don't register, if you don't disclose, there are 15 criminal offences created by that Act.

You see, regulation creates criminal offences.
So that is the injury of the Data Protection Act.
And then comes the insult added to that injury - you will pay for this!
Every data user in the country who is registered has to pay for it.
It is a system of taxation of business in this country.
Regulation and privacy is based on the premise that since some people may be wrong, everyone must be supervised and regulated - that's really the main thing I dislike about regulation like the Data Protection Act.
It says that because some people may abuse personal data then everyone must be supervised, inspected, registered and they must all disclose.

AUDIENCE MEMBER

Can you define 'privacy' for us?

BRIAN NIBLET

[handwritten note: Server privacy — secure access]

But you can't define privacy.
Privacy is not anchored in reality.
It's not something you can create a general right about.
The Common Law does not develop abstractly a general right of privacy. What it does is it supplies a bundle of rights that are anchored in reality that deal with privacy.
I want a lot of privacy.
And, by the way privacy costs money.
No-one is going to put a CCTV in my street because I own the street.
I live in a private road.
And with my neighbours we protect our privacy in our private road.
Anyone who enters the road is a trespasser.
So I'm prepared to pay money for privacy.

Text on screen: **TRESPASS**

If you assume every person has a right to his own person, has property in his own person, they can be summed up as 'property rights'.
The great action for trespass – a very important protection of privacy - used to be called 'breaching' or 'breaking the close'.
That's what trespass is - crossing a barrier, a civil remedy for crossing a barrier.

Thats what privacy is about.

Trespass is a great way of protecting privacy.

Trespass to land, trespass to the person, trespass to chattels.

The action for nuisance, which is accompanied by possession of land, deals with the peeping-tom type of infringement of privacy, property in one's reputation – protecting against slander and libel.

The action for breach of confidence, which has been developed pretty well by the judges, is a great way the Common Law supplies for protecting the privacy of confidential information.

Then there is the law of contract, the private treaty between 2 parties.

I'm demonstrating today, that can protect privacy.

Particularly the employment contract.

The contract for disclosure of information.

The contract to a user of a database.

All these things are things which the contract can protect.

Text on screen: **COPYRIGHT IS A GREAT LAW**

Copyright Law can be used to protect information.

Now, as part of the contract, I own the copyright to the talk I am giving.

My privacy is protected today by copyright.

Copyright is a great law.

Of course, it's now enacted in statute most recently in the 1988 Copyright Patents and Designs Act.

And it's been developed through statute.

But it's really a part of the Common Law.

It's a great property law.

I think it is the most important of our property laws.

We are moving into a world in which copyright will be the main source of wealth in the world.

It's the foundation of the publishing industry.

It's the foundation of the entertainment industry.

It supports the computer industry.

Programmes and databases are protected by copyright law.

The richest man in the USA, Bill Gates, owes his wealth to copyrighting computer programmes and that was a genre of work that didn't exist 40 years ago when he was born.

A great property right, copyright.

And, it protects privacy!

Text on screen: **WE ARE GIVEN MORE AND MORE PRIVACY**

We didn't know that the computer programme was going to be a literary work. But this demonstrates how the copyright concept is open-ended, these concepts can be developed.
Now I want to make 2 points about the internet.
The communication of digital messages by encryption on the internet is a major historical advance in privacy.
We've heard this morning about the mathematical discovery of the Public Key Encryptor system.
Quite cheap apparatus, with practically no way of intercepting or discovering the content of a message and authenticating the identity of the sender or the receiver.
That's a major historical advance in privacy.
By the way, the history of the world can be summarised in one sentence - the history of increases in privacy.
Primitive tribes didn't have privacy.
We are given more and more privacy, and the internet is a major advance in that.
What should be applied to the Internet is the Common Law principle enunciated in the case of Entick v Carrington 1765.
Over 200 years ago, the Kings Bench Court said that a judicial warrant is necessary to pry into a subject's personal papers.
You have to have valid suspicion of wrongdoing, and then you have to have a judicial warrant.
Then you can pry into the nature of the message, but not otherwise.
Because the court said "For papers are often the dearest property a man may have."
But in the 21st century electronic communications are going to be the dearest property a man may have.
And I say that principle, which is a Common Law principle, is the one to apply to interception of messages on global electronic networks.

Text on screen: **NOBODY OWNS THE INTERNET**

People say that the internet is a great challenge to the Law.

Margin note: Contradiction between server as Commons and as Bureaucracy)

They say, intellectual property is ended because of the internet, everyone can defame everyone else.

Absolutely false.

The trouble with the internet is that it is not a proprietary system.

The internet at the moment is an inchoate feudal system.

Nobody owns the internet.

No-one administers the internet.

No-one manages the internet.

You can't buy the internet, you can't sell the internet.

Propriety doesn't come into it.

Its common land.

In the old days if a man put his cow on common land of course someone would come in the middle of the night and milk it.

Because it was not a trespass to do it.

You can't apply legal principles in a framework of lawlessness.

Thats the trouble with the Internet.

But we see the emergence of proprietary nets, intranets, and that will change the situation.

You see in the middle ages when we had this feudal common land, when the Lord of the manor gave all this common land to his surfs to put cows on it, he kept his land separate from them.

This is what is happening at the moment.

We get the corporate intranets with their firewall to the common land.

Thats how they're protecting it.

There's no infringement of the intranets, that's too well-managed.

If there were, Common Law principles would apply.

And then there were the great Enclosure Acts of the later middle ages, the privatisation of land.

We move from the feudal system to the capitalist system.

Land could be bought and sold for the first time.

Enclosure acts on the common land, privatised the land, and that is happening on the internet.

The Internet will emerge from proprietary networks.

In the century that followed the Enclosure of common land, agricultural productivity accelerated within a hundred years.

A lot of trouble ensued over the Enclosure Acts as you know, social upheaval.

But in that hundred years afterwards, agricultural productivity increased by a factor of five or even maybe ten and resulted in a great increase in population

because they could be fed for the first time.
And I say that the emergence of proprietary digital global networks, where people can buy and sell and administer and manage, will solve the problems, and we shall see ten years after - it won't take a hundred years - an enormous increase in international commerce on the internet, we are waiting for that.
I wouldn't put a book on the internet now.
But on a proprietary network of course I would be pleased to put a book on the internet because I know that legal principles can be applied to a maintained and managed proprietary system.

Text on screen: **NATURE TO BE COMMANDED MUST BE OBEYED**

So now, Chairman, I'm in a position to answer the question that is the subject of my talk which is - engineering and law - are they compatible or contradictory?
In answer I want to quote Sir Francis Bacon, who was a distinguished scientist and a distinguished common lawyer. I think he understood the nature of law. What he said was "Nature to be commanded must be obeyed".
An engineer is a creator, he creates structures, he creates things out of natural materials and materials derived from Nature.
An engineer, to command his structures, must obey nature.
The Common Law is this aggregate of commands based on nature.
They are commands and in order to be effective, the Common Law commands must obey nature.
So of course engineering and the Common Law are compatible, they are anchored in reality.
But engineering and regulation may not be compatible.

Everyone Claps

CHAIRMAN
Thank you Brian. If you have any specific questions for Brian please go ahead.

AUDIENCE MEMBER
So if one of us in the audience had recorded this, what would the position be then?

REF: IR0006
NAME: VoIP PBX
TYPE: TRANSCRIPT from teleconference.
DATE: 2013

- Demonstrate use of noise as material meeting place. The
- Discussion laborious suffering from substantial echo.
- Mention of the link archive and making DVD copies. No evidence.

+ meeting software VOIP PBX mgt called Asterix PBX, a technology reminiscent of the "cone of silence" from the 1960's television spy series "GET SMART". The cold war aesthetics of which some members may have imprinted as childhood. The "cone" was an iconic device designed to protect conversations from security subs by enclosing the users in a transparent sound-proof shield, which being substantially defective delivered more interference than signal.

+ the counter-efficiency of the Irrational PBX as a telecomf medium - effectively a regression to the analog conf. call, an unwieldy pre-inet techn. where remote discussants in the previous century were forced to meet - IS POTENTIALLY CONSISTENT WITH THE IRRATIONAL GROUP'S CONTESTATIONAL WORK ETHICS &

R. IR0006/01 Meeting 2

Mon, 23 Sep 2013 21:01:13 +0100

$server*CLI> WELCOME TO IRAYYYTIONAL DOT ORG CONFERENCE ROOM. PLEASE DIAL YOUR REQUIRED EXTENSION.
$ **sysop** (DIALS 600)
$ **sysop** (ENTERS)
$ **thief** … it's pretty echoey in here.
$ **treasurer** is that you secretary?
$ **gardener** it's gardener...
$ **treasurer** oh. hi gardener!
$ **gardener** hello. hi treasurer. is that thief as well?
$ secretary (VIA EMAIL) i can hear treasurer and sysop and a massive echo, but no-one can hear me
$ **thief** hi! how're you going?
$ **gardener** ok. pretty tired. all good.
$ **treasurer** how are you calling us, gardener?
$ **gardener** i'm calling on a landline but it's so echoey i don't know if i'm going to be able to... it's like a psychedelic song.
$ **treasurer** you get used to it after about 10 minutes.
$ **gardener** there's loads of feedback.
$ **thief** yeah it's kind of fun! (kind of fun kind of fun)
$ **gardener** ah! ok. who else is in the cave?
$ **treasurer** sysop's in here somewhere.
$ **gardener** are you there sysop?
$ **sysop** yes.
$ **gardener** hello.
$ treasurer (VIA EMAIL) cleaning lady, we have a budget! how about you put a few euros on skype & skype in to the conference room.
$ secretary (VIA EMAIL) i'm using mac and x-lite ... does thief have any tips on microphone?
i mean, no-one can hear me in this spaceship.
$ **treasurer** which software you using sysop?
$ **sysop** twinkle.
$ **treasurer** cool ok.
$ **gardener** i'm using a landline. i'm in henley, just outside oxford.

$ **treasurer** cool. we're just trying to fix secretary's microphone and get cleaning lady online
$ **gardener** uhuh. thief, how've you been?
$ **thief** good! are you coming to xxxxxxxx?
$ **gardener** yes, are you?
$ **thief** i hope to. it depends whether, you know, it makes sense.
$ **gardener** that old thing. whether it makes sense or not.
$ **thief** if everyone else is coming, yes i think so.
$ **gardener** good. oh. there's some ringing going on.
$ cleaning_lady (VIA EMAIL) i was trying to buy earlier (from my money) . but there is a veryfication code coming to my spanish mobilephone associated to my bank account, mobilephone is not in use
$ treasurer (VIA EMAIL) can you use someone elses skype account with money on it ? xxx
$ **treasurer** so gardener, i've got some money to give you.
$ **gardener** oh?
$ **treasurer** yes it turns out we're both in a show. i thought it was just me. bordercrossing slideshow.
$ **gardener** oh no not again...
$ **treasurer** so haven't you got a computer any more?
$ **gardener** i have, i just thought calling by a landline would be easier.
$ **treasurer** yes?
$ **gardener** but it's really difficult to hear what's going on. anyway...
$ **treasurer** who's going to lead this meeting anyway - cleaning lady wasn't it?
$ **gardener** what's the plan? are we going to talk about what we're going to do?
$ **treasurer** sysop can you lead this meeting?
$ sysop ok...
$ **gardener** it's a strange meeting.
$ **cleaning_**lady (ENTERS)
$ **treasurer** hi cleaning_lady.
$ **cleaning_lady** i can't hear very well. why don't we talk on

e-talk?

$ treasurer (LAUGHS)

$ cleaning_lady i mean it's really difficult to understand what you're saying.

$ treasurer this is our meeting! this is the work.

$ cleaning_lady it's very challenging...

$ treasurer so let's make the work right now, it's an audio book and we're talking about what we're going to do.

$ cleaning_lady ah you're talking about the book...

$ treasurer sysop's recording this.

$ cleaning_lady where's thief?

$ thief i'm here.

$ cleaning_lady ah you said nothing! are you hiding behind the bushes?

$ thief no! to tell you the truth i'm just a bit worried about the echo, i'm trying to eat almonds in the quietest possible way.

$ cleaning_lady echo? (echo, echo)

$ treasurer (LAUGHS)

$ cleaning_lady ok. who is moderating?

$ treasurer sysop is, i think.

$ cleaning_lady ok. so here i am, we can start the meeting. this is not serious... this phonecall is going on irational's bill, so make use of it.

$ secretary (ENTERS)

$ treasurer hi who's that?

$ secretary it's me! secretary. sounds amazing in here. Wooooooo..... (wooo wooo wooo)

$ treasurer great, we've got everybody.

$ secretary so shall we talk about the xxxxxxx software? did anyone see it at xxxxxxxx?

$ cleaning_lady secretary you have too much echo. i can't understand a word you're saying.

$ secretary i don't know what to do...

$ cleaning_lady (LOUDLY) speak more silent, you're too loud!

$ sysop (LAUGHS)

$ secretary (WHISPERS) is this better?

$ treasurer i have a suggestion. is everybody using

headphones?
$ **gardener** i can't hear properly.
$ **secretary** i have a question. is it possible (possible possible) to access the disk remotely?
 (remotely remotely remotely)
$ **treasurer** access the disk remotely? yes it is possible. i could send you a copy if you want.
$ **secretary** on a drive?
$ **treasurer** on a drive or on a dvd.
$ **secretary** ok. great
$ **cleaning_lady** what?
$ **treasurer** secretary wants a copy of the disk. the archive.
$ **cleaning_lady** ah ok.
$ **treasurer** maybe I could just send everyone a copy on a dvd.
$ **cleaning_lady** on a dvd... but how big is it?
$ **treasurer** it's probably not that big. i can try to cut out things that aren't necessary. all send me your postal addresses and i'll try and send you copies...
(THE CONVERSATION CONTINUES. OVER STRATEGIC AND MATERIAL THINGS, UNTIL IT IS TIME TO WRAP UP)
$ **secretary** did you get any recordings sysop?
$ **sysop** yes.
$ **cleaning_lady** i did.
$ **secretary** i think this should be our exhibition. just the audio.
$ **cleaning_lady** ok. what more?
$ **treasurer** that's all there is, cleaning lady.
$ **cleaning_lady** so the meeting is finished. maybe. you're staying there because you like to hear this electronic wind!
$ **treasurer** (LAUGHS)
$ **cleaning_lady** you're hypnotised!
$ **treasurer** let's sit in silence and listen to the wind together.
$ **cleaning_lady** waiting for all the people to get silent. everybody silent please. ok.
(SHORT SILENCE OF ECHOES)
$ **anon** schhwwooo
$ **anon** whhheeeoooooo

```
$ anon  whahahahhacchhhhyyyyyyeee
$ anon  ttsssyyyyyy!!!!yyy!!!!
$ anon   qq!! qqqww!!! qwww!!! qqqqqwwwY^Y^Y^!!!
$ treasurer  you wouldn't get this on skype would you?
$ cleaning_lady  i have my dog here he is looking at me.
$ secretary  get him to join in!
$ cleaning_lady  he is not interested.
$ treasurer  where's the moderator? come on, call an end to the meeting.
$ thief  good night y'all.
$ thief  (LEAVES)
$ gardener  (LEAVES)
$ cleaning_lady  lady bye.
$ treasurer  (LEAVES)
$ treasurer  bye.
$ treasurer  (LEAVES)
$ cleaning_lady  see you soon.
$ cleaning_lady  (LEAVES)
$ sysop  (LEAVES)

server*CLI> sip show peers
11 sip peers [Monitored: 1 online, 0 offline Unmonitored: 10 online, 0 offline]
server*CLI>
```

```
Sun, 17 Nov 2013 19:15:24 +0000


$server*CLI> WELCOME TO IRAYYYTIONAL DOT ORG CONFERENCE ROOM.
PLEASE DIAL YOUR REQUIRED EXTENSION.
$ sysop (DIALS 600)
$ sysop (ENTERS)
$ imposter ...i'll give you back to treasurer. i don't know
what you're saying (((saying saying saying)))
$ treasurer hi there (((((hi there hi there hi there hi there
hi there hi there hi there)))))
$ cleaning_lady say something (((((say something say
somethingsay something say something)))))
$ treasurer $cleaning_lady can you use your headphones
((((((headphones headphones headphones headphones headphones
headphones))))))
$ treasurer have you got headphones (((((((have you got
headphones have you got headphones have you got headphones
have you got headphones have you got headphones have
you got headphones have you got headphones have you got
headphones)))))))
$ cleaning_lady there was no echo before ((((((((there was no
echo there was no echo there was no echo there was no echo
there was no echo there was no echo there was no echo there
was no echo))))))))
$ treasurer you have to use the headphones (((((use the
headphones use the headphones use the headphones))))) can
you ((((can you can you can you)))) find some headphones
$cleaning_lady? ((((headphones cleaning lady headphones))))
$ cleaning_lady yes I can ((((yes yes yes yes yes yes)))))
$ cleaning_lady (TYPES) ((((((types) (types) (types) (types)
(types) (types) (types) (types) (types))))))
$ secretary (VIA EMAIL) i can hear everyone, but no-one can
hear me...
$ cleaning_lady are you there $sysop? (((((($sysop $sysop
$sysop $sysop $sysop $sysop $sysop)))))))
$ cleaning_lady hello ((((((hello hello hello hello hello
hello hello))))))
(LONG SILENCE OF ECHOES)
```

$ treasurer is that you $cleaning_lady?

$ cleaning_lady yes.

$ treasurer can you hear me now?

$ cleaning_lady yes.

$ cleaning_lady ok ((ok)) what is going on? ((what is going on?))

$ treasurer (VIA EMAIL) twinkle is working!

$ treasurer someone's coming in... (((someone's coming in someone's coming in someone's coming in))) via a telephone (((via a telephone via a telephone via a telephone)))

$ secretary (ENTERS)

$ secretary hello (((((hello hello hello hello hello hello hello hello))))

$ treasurer hi $secretary ((((((hi $secretary hi $secretary hi $scretary hi $secretary hi $secretary hi$secretary hi $secretary))))))

$ secretary it must be me ((((((be me be me be me be me be me be me))))))

$ secretary that echoes ((((((that echoes that echoes that echoes that echoes that echoes))))))

$ secretary i don't know how to ((((((i don't know how to i don't know how to i don't know how to)))))))

$ secretary get rid of it ((((((get rid of it get rid of it get rid of it))))))

$ treasurer ...headphones ((((headphones headphones headphones))))

$ secretary hello?

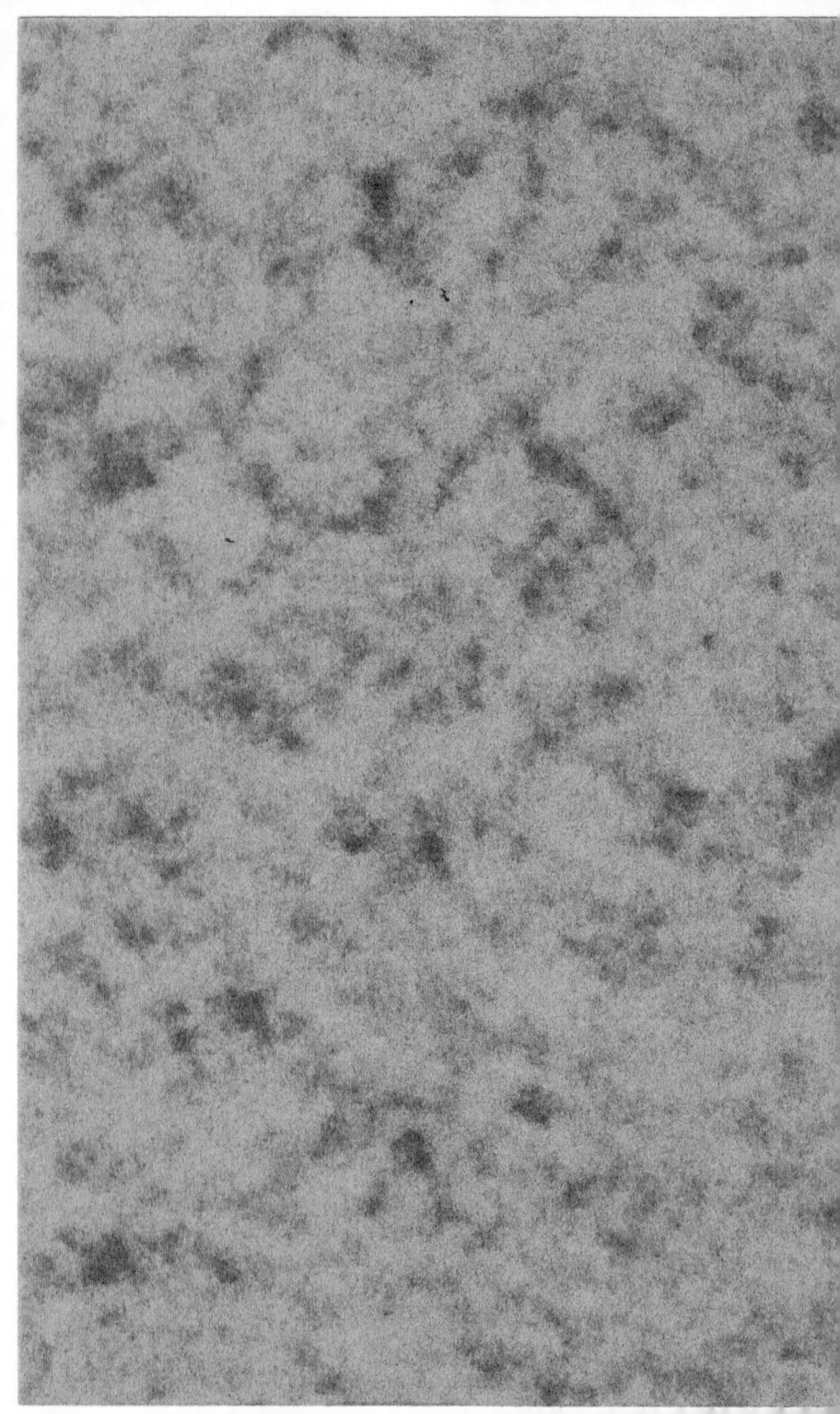

IRATIONAL PARK

Ref: IR0013, IR0014, IR0015
Name: IRATIONAL PARK PLAN
Doc type: 1 x plan visual, 3 x plan indexes
Date: December 2013

This sequence of material provides evidence that a park was being prototyped. It is not known whether the park was actual or imagined, but these sketches appear as preliminary strategies for renovation and anchoring of the irrational dink in biological form. I speculate that the meetings in Lüneburg were to plan the migration of the newer data onto a physical park.

Could some dink elements have been 'encoded' into the park as a means of secure data dispersal? The abandoned park at Lüneburg where I found the plaques is most likely spot.

IRATIONAL PARK - SKETCH 01

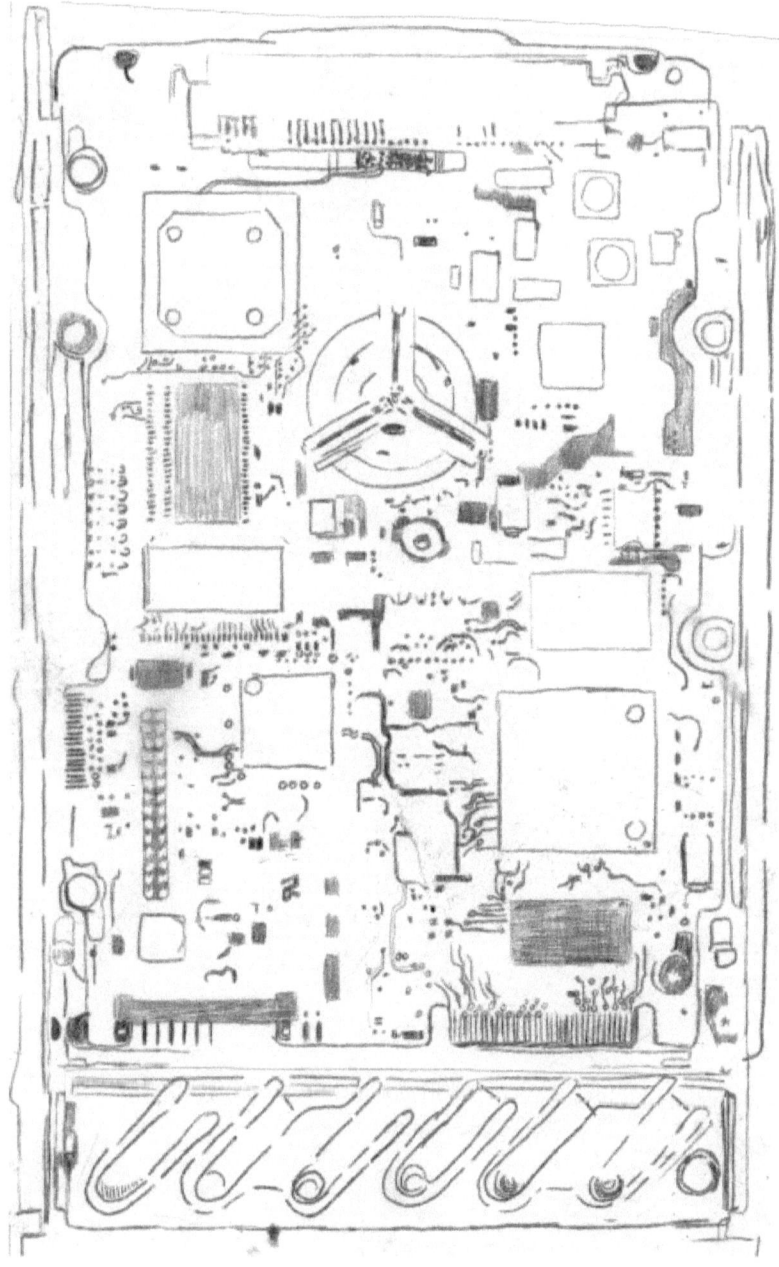

PARK IRATIONAL - MANIFESTATION OF COMPOSITIONAL SENARIOS 01

Irational Server Park (ISP) - Compositional Manifestation 01

Draft 01 Relational and spatial scenarios for ISP.

SITE

13. Forest
12. Tidal River, with no opportunities for tourism.
10. Disused single gauge train line. Community pressure group 'friends of the line' hope it will one day run again.
6. Irational Server Park holds a 130-year lease on 16 acres of land, comprising a post-industrial wasteland and wayside woodland.
7. The M7
5. Last bus stop for the number 212.
2. Family run newsagents.
11. Housing estate built in 2011 by 'Nectar' under the part buy part ownership scheme.
5. 24 hour supermarket and petrol station.
1. Leaving their den in ISP, Foxes travel here to raid the Waitrose bins.
8. Hospital.
3. From his house Jason (the thief) often walks to the park, the other night he stopped by the newsagents for a lighter.
4. Sports Academy opened the same year as the park, soil from Irational Server Park is collected into a locker by the artist.
14. Highest point in City. From here you can see ISP, like an empty haze of green.
15. The city's main station. In the early hours a procession of people walk toward it.

TREES

9. Fig grows here, leaning along the southern face of control tower planted in hope of fruit and concealment.

8. From this point till the edge of the park runs an uninterrupted line of Lime trees. Planted by keeper 5, to supply irational users with rope and sandals. Lime, a symbol of liberty, a twin to data artefact 'borderxing'. On the eve of May the flowers are soaked and drunk and the branches are used as dancing sticks.

11. Mulberry tended by both koopers 8 and 3, their vested interest often resulting in dispute.

14. Yew predates park irational by 1,000 years this sacred gnarled symbol of death and transcendence with its toxic leaf and seed, yet edible bright red arils. Twin to irational's digital form and habitually used for all park annual general meetings.

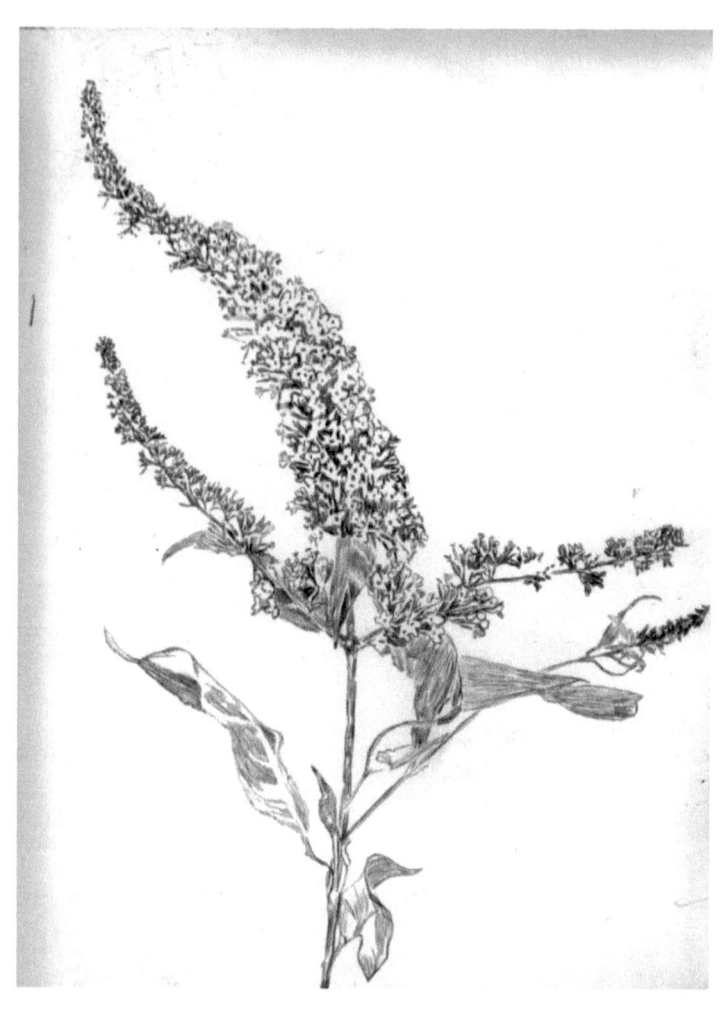

CUT THE SCREEN
CRACK THE LENS

7. Silver Birch twinned with 'cgi-bin', used by all keepers to make broomsticks, wine and paper.

5. A Hawthorn hedge circles zone 3 planted by keeper 5. Hawthorn defines the zone's edge and her personality. A plant of great resistance with hard branches full of knots thorns 'Zdrav ko dren' they say. The hedge is home for Sparrors and destination to visitors who eat the leaves, even the Wild goats have been seen here.

12. Crab Apple, although often mistaken for the wild crab apple this tree is feral, springing to life after a core was thrown from a window of a passing train. The fruit is seasonally used in the park's cafe.

1. Balcanic Spruce Pinacea Balcanica derived from Balcan Peninsula stands here, providing shade for the volunteer run library.

4. One of countless Buddleia. Early on Buddleia became Irational's principle symbol. A sign next to it reads CUT THE SCREEN CRACK THE LENS. Loved by butterflies.

2. A collective orchard, a place used for sleep and branches used for spoons.

6. Hazel grows in abundance, riddled with a network of aerial platforms, walkways and paths frequented by all of the keepers and squirrels. Keeper 4 feels that Hazel represents an essential part of his outer shell, keeper 5 makes crochet needles.

10. Cottonwood indigenous to North America, a sapling carried in a suitcase and planted by friend and user of irational twin to work 'ndnnrkey'.

14 Ginko, a living fossil, its leaves are gathered to make a tincture for neurological ailments (keeper 6 trades it for batteries).

AMENITIES

1. PBX Cave Site - An artificial cave constructed by keepers 3 and 2, a lovely cave decorated with shells and flint. For park users to gather, commune, sing, using the cave walls to create acoustic echo.

10. Routeless - A disused single gauge train line that leads nowhere for imagining and plotting journeys. A starting point for hitch-hiking.

6. Chalk quarry - Disused, for sourcing chalk and throwing stones.

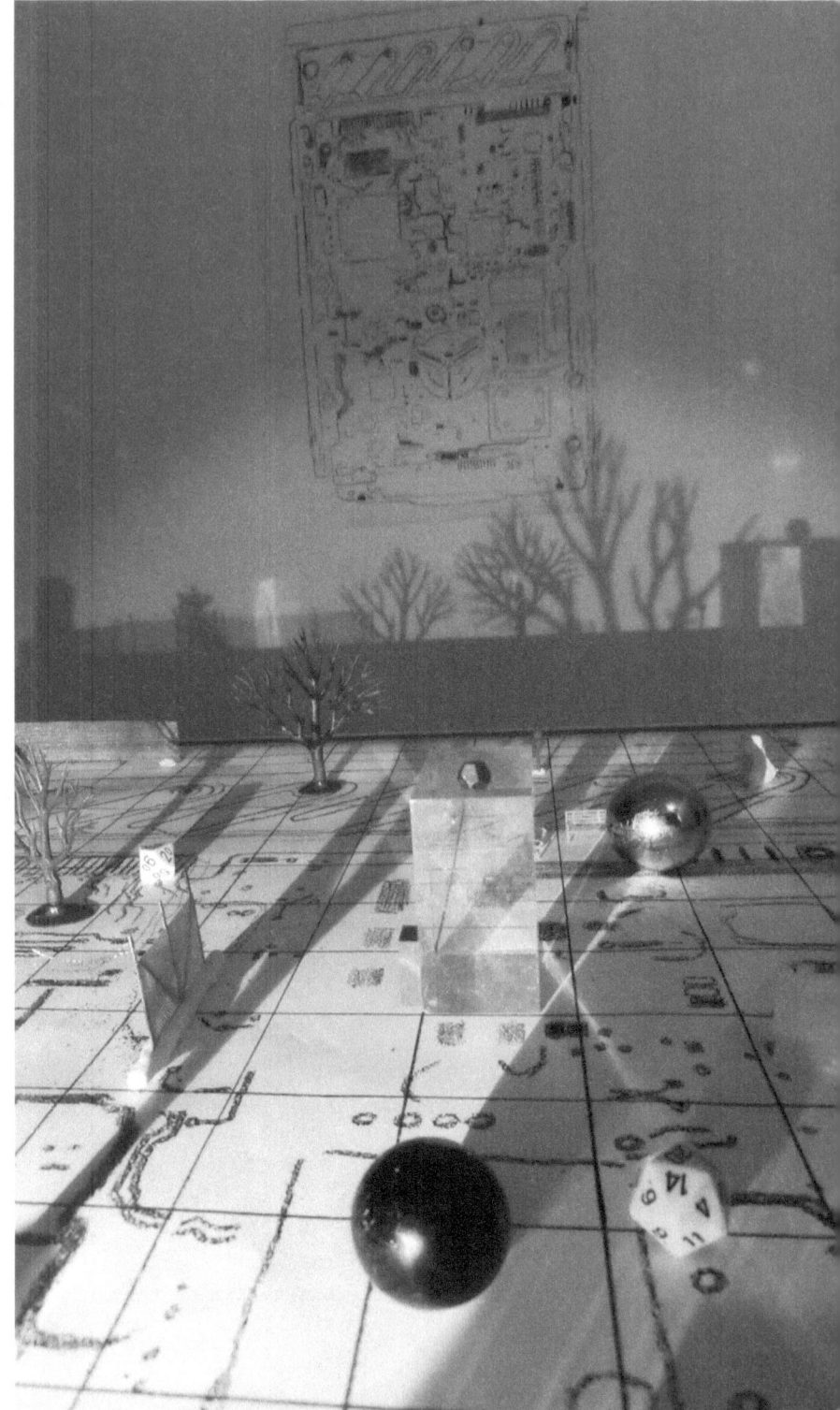

0. Tour de Fence - An area of climbing apparatus structures made of wire, steel, copper and brick, for training to overcome vertical defensive barriers

5. ssh Chicken run - Two hen coops separated by a series of wire runs. Provides communal fresh eggs. Secured to keep out the foxes.

2. COURIER NEWS - A newsagents, parcel collection and PO Box facility

7. FERAL Cyber Cafe - open 24 hour.

3. Cgi Bin - Shed for tools to maintain the functionality of the park

12. Search.pl - A watchtower observation point and hawks' nesting place. It's difficult to climb but from here you have a birds-eye view of Irational Park.

9. Tunnel entrance for discrete departure.

4. Vending Machine - Sells 8 different types of soft drink, currently out of order due to vandalism.

8. Lab - used to make experimental produce, contributes to 40% of the parks income.

1. Libary - volunteer-run, open seven days a week.

12. Phone box next to the Potato beds.

PARK AT MIDNIGHT

11. Hole in the fence hidden by patch of nettles.
1. Tree laden with walnuts.
8. Empty park bench.
4. Mice squatting the vending machine.
9. Zone 6 in the fields of entropy eyes glisten.
4. Unemployed park attendant rolls a cigarette.
12. Far corner of zone 6 - 'the void' there is some laughter.
14. Old homeless woman with her trolley there she is again.
12. Man drinking with friends.
7. Fallen down tree dark in shadow.
8. Lights coming from lab window.
2. Hedgehog scurries past.
6. Fire pit lights up silent faces.

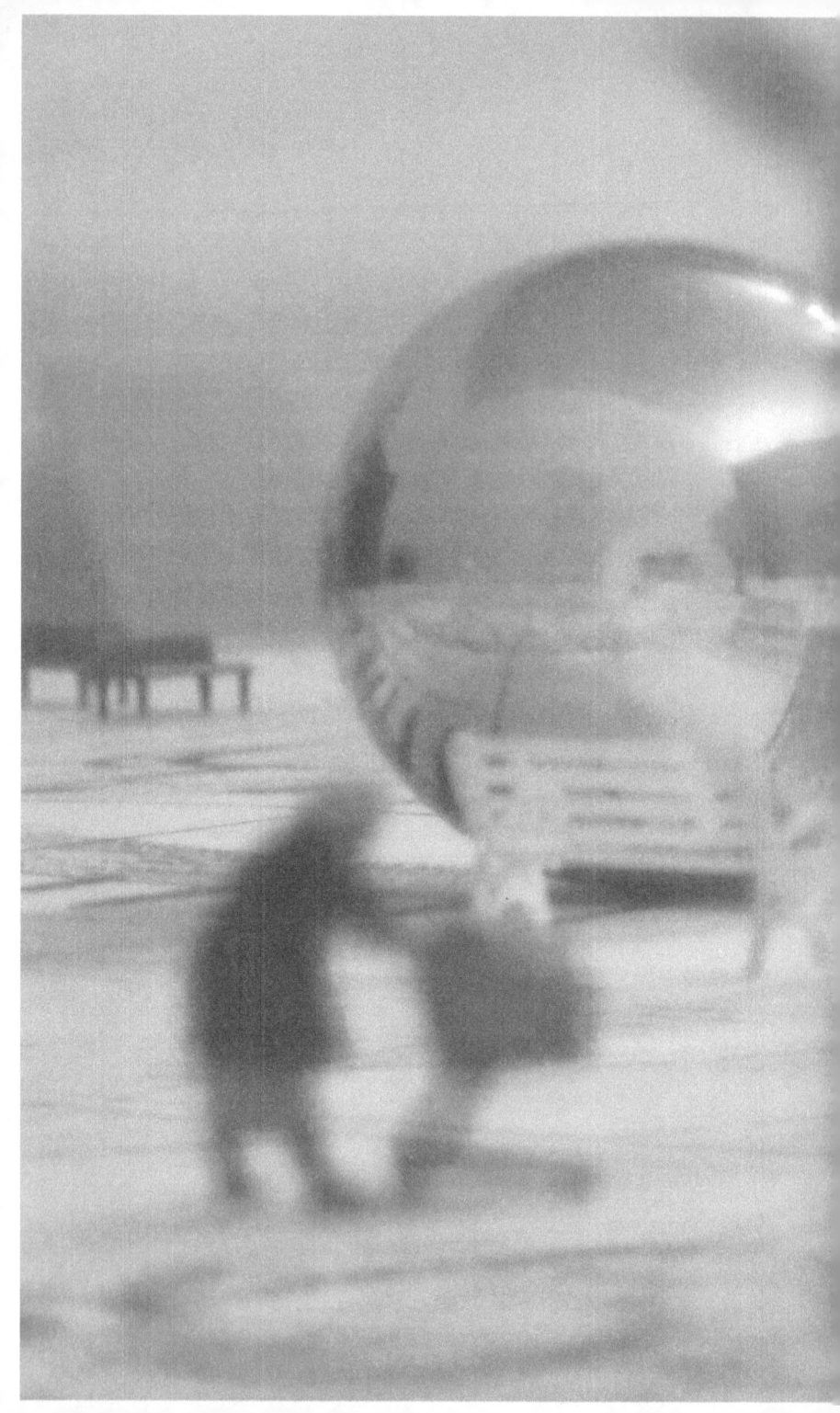

Ref: IR0026
Name: IRATIONAL PARK?
Doc Type: PHOTOGRAPHY (V.M.)
Date: February 2044

Now overgrown, I believe this abandoned parkland located on the edges of Tinsburg Heide was to be the site of Irational Park. The city owned the land but they hold no records of Irational contractor. This picture was taken from a tower, possibly the search-Pl. observation point.

If the Park was ever realised, it was unofficially developed, an nichet, but in plain view.

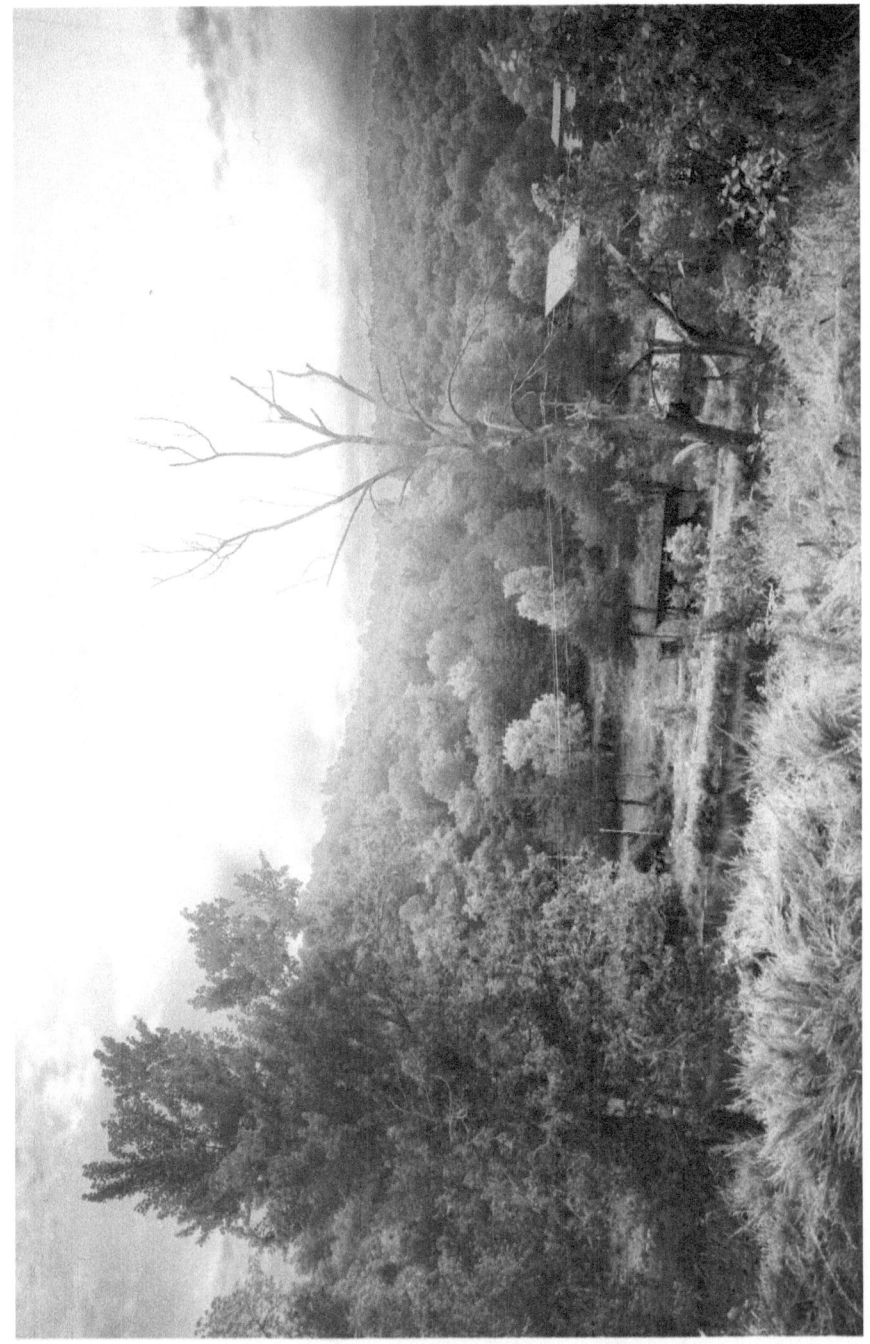

Ref: IR0D25
Name: PARK PLAQUES. DISPUTED HISTORIES OF IRATIONAL.ORG?
On Topic: Photo and reconstruction of writings from the plaques.
Date: February 2044

On the eastern side of an abandoned park in Lüneburg were found 8 metal boards embedded in the big rock monument. I assume they served as a kind of information plaques for the Irational.org Verver Park. Damage by rain and corrosion make the engraved text barely readable. Signs of intentional damage and erasure. After cleaning I managed to extract some consistent fragments of information on 4 of them. Content was sometimes repetitive, divergent, contradictory. The plaques seem to be histories of I.org preserved from different angles of the members.

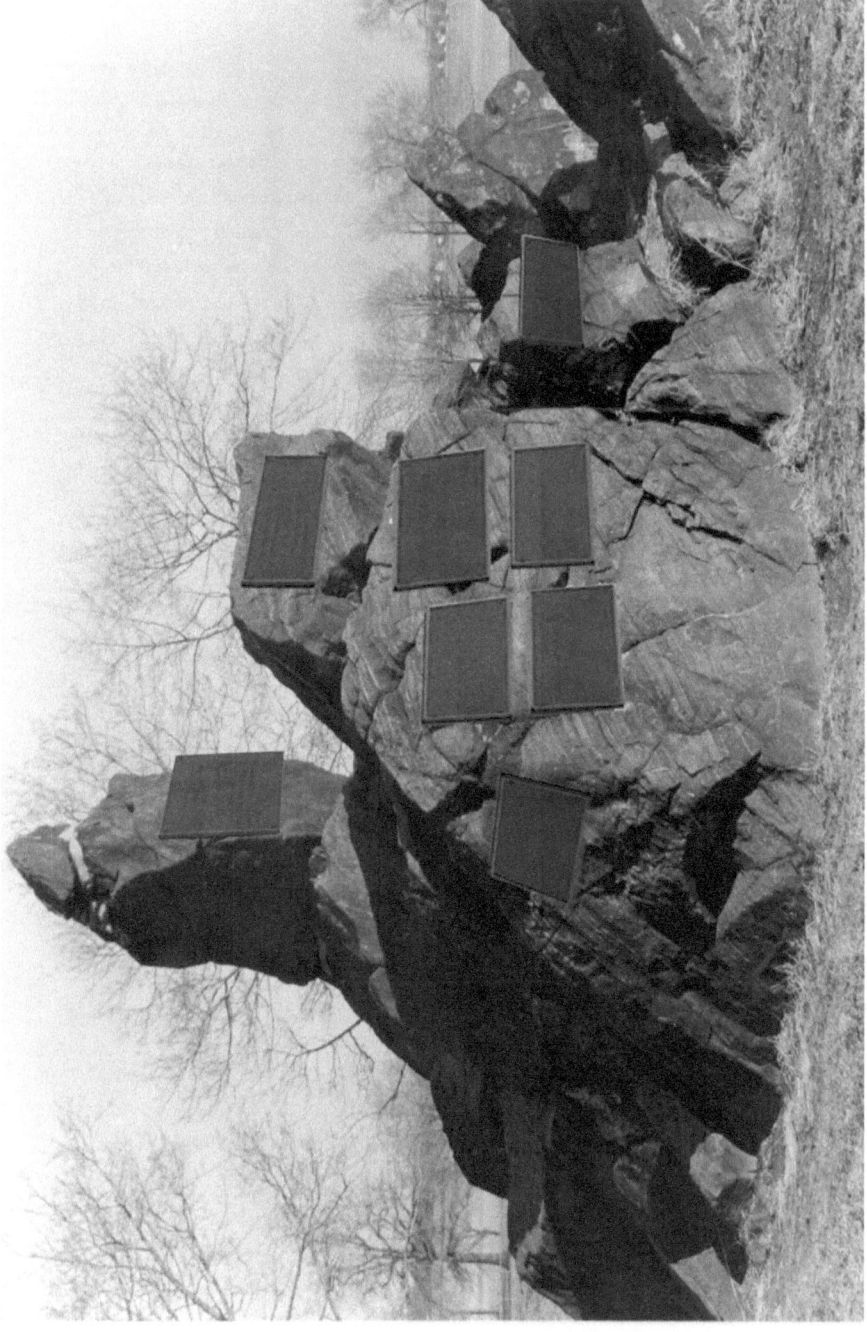

Document R.IR0025/02
Reconstruction of the writings from the plaques

Plaque#1
*.... byproduct .... open a euros-denominated bank
account... ..... failed). ...... corporate form - ....
exoskeleton ...... the individuals ..... trade, interface,
be communicated, narrativised and amplified collectively to
outside entities .... art collectives, curators, banks
and The Law) .... server, t..... ... mailing list
and the Asterix PBX,.... .... to 'meet'...... ....
.... to-and-fro, ...ideological struggles .....
violent fights..... ....compiled ..... internal records .....
affection, friction .... devastation ..... group..... ......
.... times clinging to rafts.... .... data dump..... future
historians .....*

Plaque#2
*..... .... cafe BBS (Bulletin Board System) in
1995..... .... technical ap..... .... ystem donated by
Ivan Pope..... ..... ternet doma....... ....
early experimentation... art ...cultural practices .....
90's London.... .... i286 PC with a Microsoft DOS .....
Wildcat bbs soft..... . ... with technical support
f..... ....500 transient .... ces,
politics and desires..... offer a user account ....
relationship...... dial into the system ....exchange
files..... messages ... . do-it-yourself .... radio/tv/
te..... es/fax/mailart/flyp...... ..... ance/computer......
domain name....... ..... old back to him.... .... a
direct..... .... es for 000.00 GB..... ad accrued
much value during t....... net cafe spa..... ....
critical and timely investme..... .... omain name
http://www.irati.....*

Plaque#3
.... 1995, ... Internet ..... introduced into the domestic sphere .... technology bubble ..... shape. .... this new utopia of freedom ..... access to information and knowledge floating .... vanish...... flowering of a new conception of power .... become immaterial ..... loss of its grounding in material resources...... witness a battle for control of knowledge .... information, a fight for it to be managed as a lucrative monopoly on distribution and circulation .... see this more clearly now....

Plaque#4
.... not a collective or an organisation. ..... name of a server used .... maintained by individual artists for .... believed ...... non-representational.......girlfriend .... official husband.... its founder. .... inner dynamics ....ignorant .... meaning of relations .... complexity ... ... professional and emotional constellation........ resignation of .... it's members .... ..are ... .. leaving ..... breaking down......

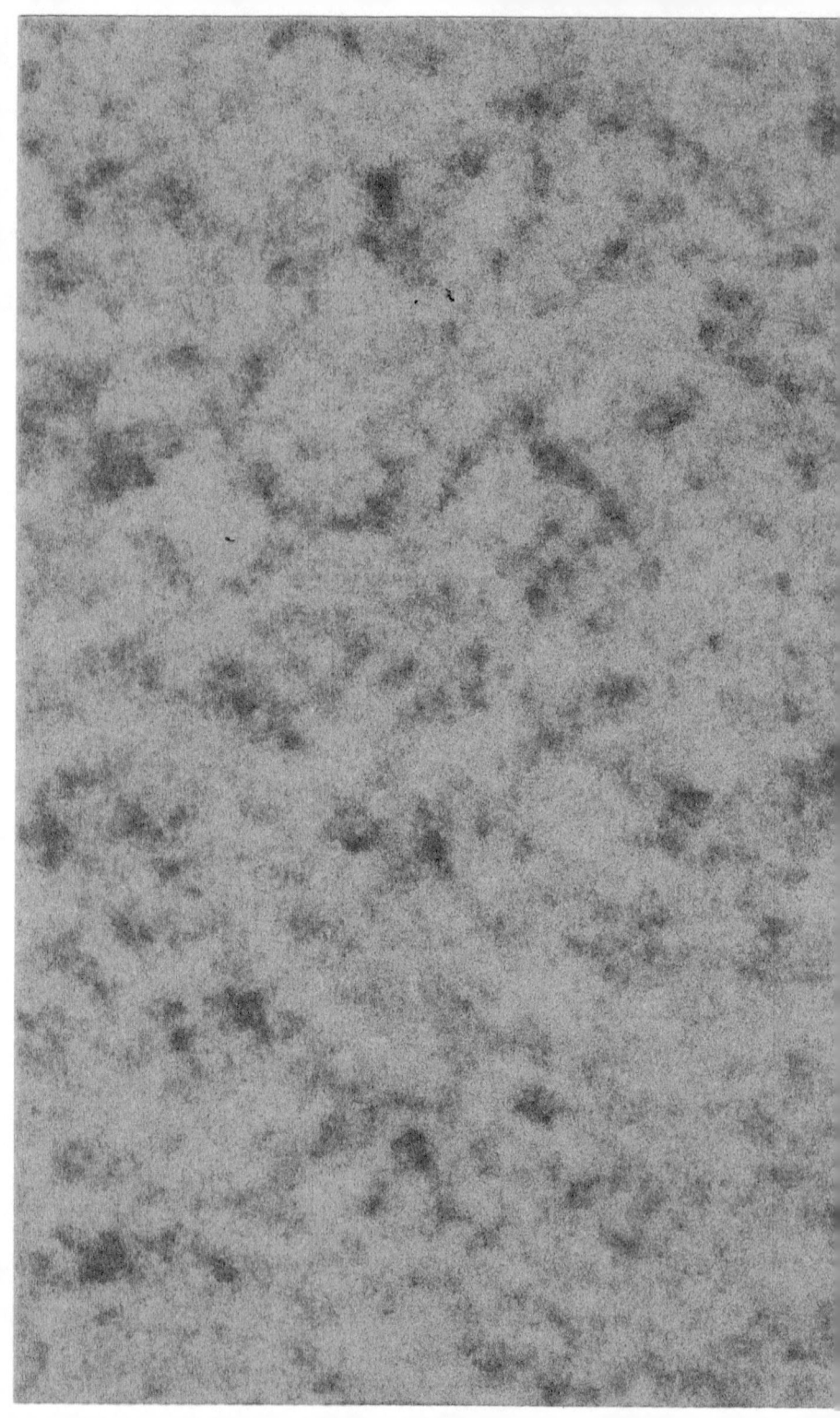

DISK
DISPOSAL

Ref.: IR0009, IR0010, IR0011, IR0012
Name: FIREHOLE
DocType: 3x photos, 1x note
Date: December 2019

A series of images depicting a plug hole. Objects for burial seem to be a fire-making kit. Were they also preparing to bury the flesh?

Symbology of the hole like fire or potent. A void, an emptiness, an inflicted wound, a passage between spatial and non-spatial.

SEE CONSTITUTION regarding relation between fire and scruer.

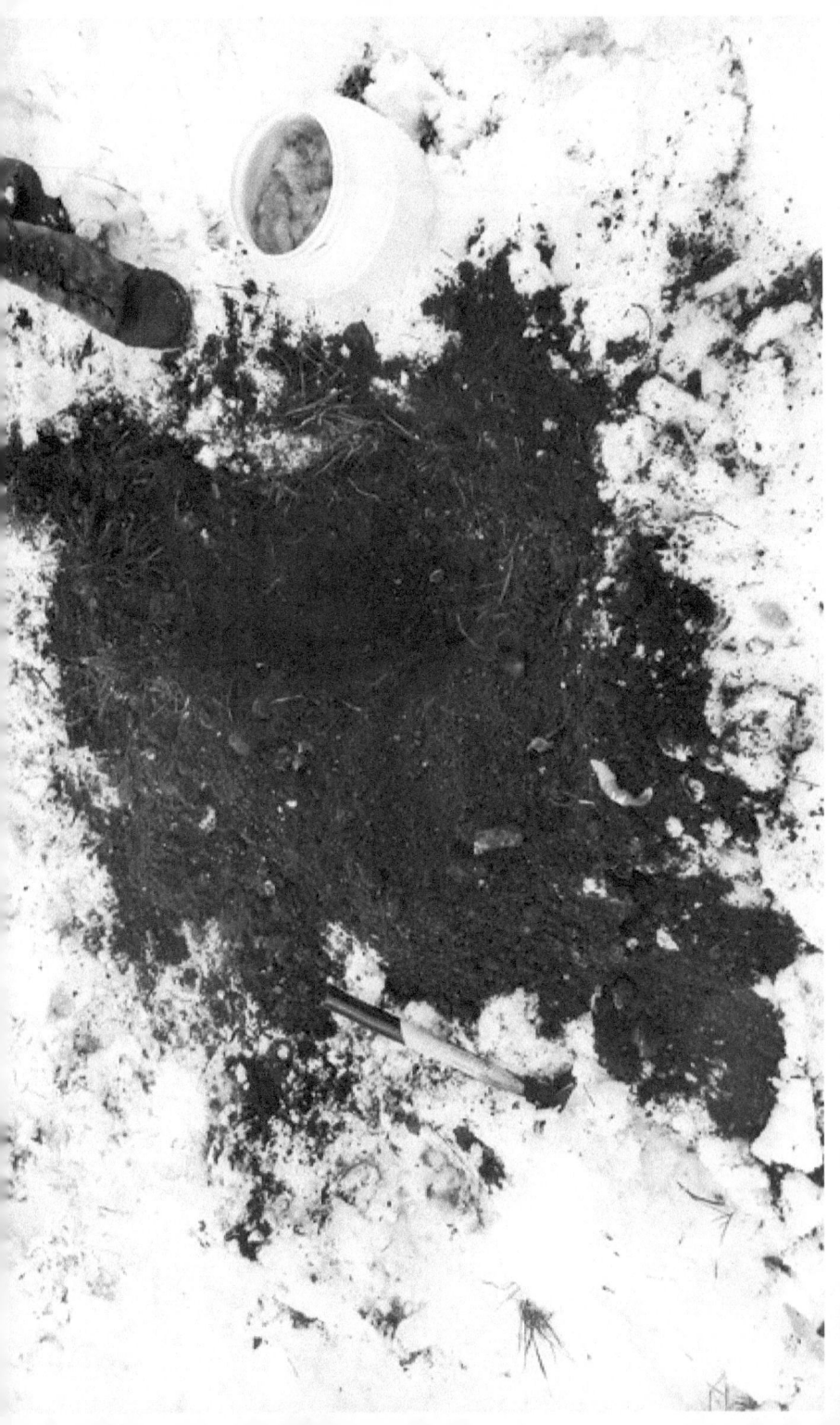

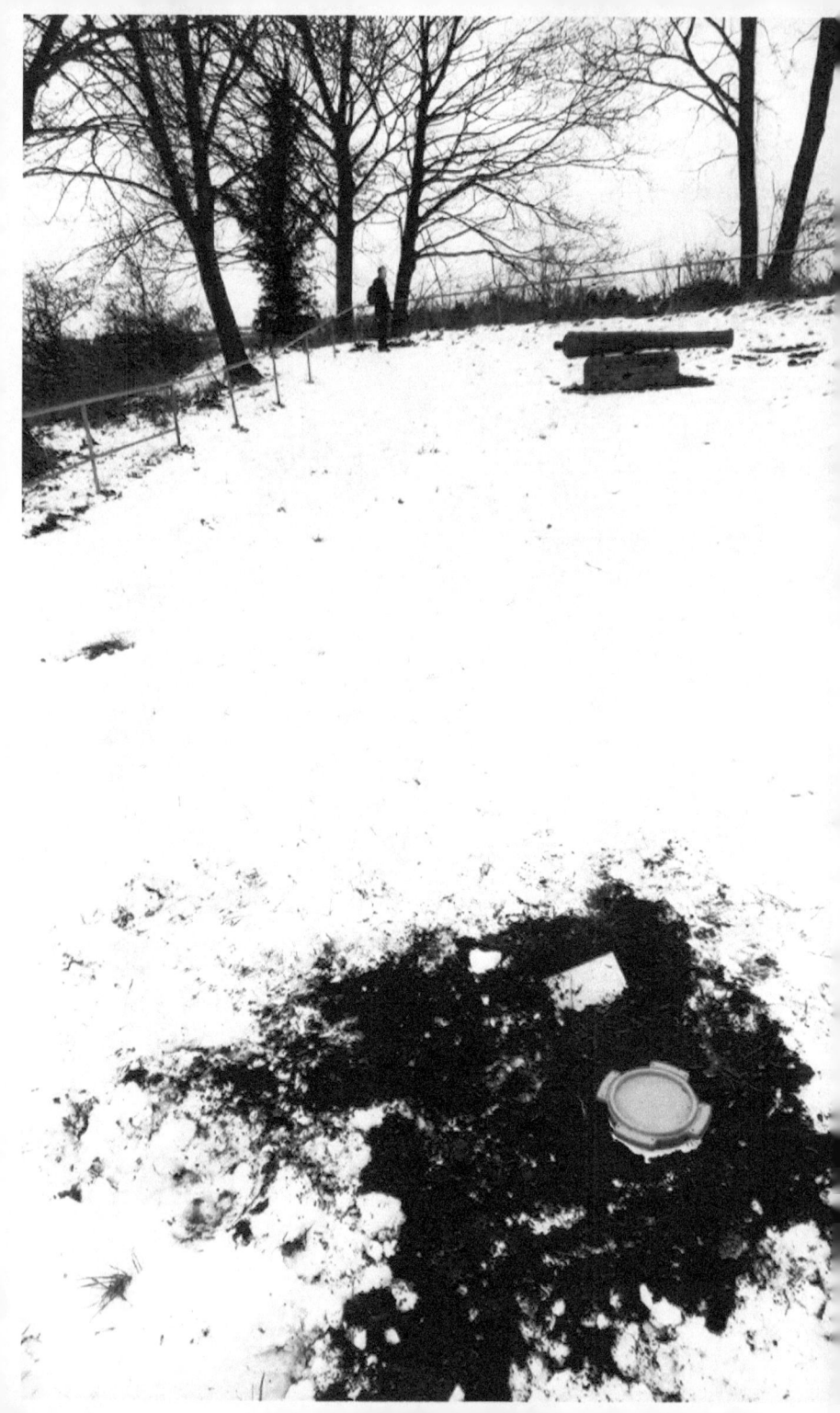

You have found a fire-making kit to make a small campfire. It's a gift from irational.org/firehole.
Enjoy! And keep warm.

Hi!
Du hast ein Feuerset für ein kleines Lagerfeuer gefunden.
Es ist ein Geschenk von:
irational.org/firehole
Genieß

THE HOLE

EXT: *The Lüneburg Heide, morning, grey mist.*
ARTIST is walking along a path, halts. The sound of a spade hitting ground can be heard in the distance. ARTIST listens then walks in the direction of the sound. Eventually ARTIST comes across GARDENER with a baby strapped to her back digging a hole. ARTIST approaches, stops to sit down on a fallen tree and watches as the pile of earth accumulates.

ARTIST *(to herself)*
Grain upon grain, one by one, until one day, suddenly, there's a heap, a little heap, the impossible heap.
(Pause)

After a few minutes she takes out a paper and pencil and starts to sketch the hole-digging scene.

ARTIST
Nice dimensions, nice proportions to this hole.

ADMINISTRATOR enters, inspects the hole-digging and addresses ARTIST

ADMINISTRATOR
Good to see some productivity around here. Whoever works the land owns the land!

ARTIST
Does that mean Gardener owns the land she has dug? (points to the pile of earth next to the hole)

ADMINISTRATOR
Actually she owes rent on it, she's working off her debt.

ARTIST
But what is the hole for? Is it a burial? Has someone died?

ADMINISTRATOR
It's an asset. A debt asset.

ARTIST
Whats it's value?

ADMINISTRATOR
Unknown. We can speculate on it. Its a speculative debt asset. The important thing is she keeps digging.

ARTIST
I thought maybe she was creating a seedbed, an incubator.

ADMINISTRATOR
No, the seedbed is creating her. In the meantime we can trade our debts, future and past.

ARTIST
I have some seeds! What shall I do with my seeds?

ADMINISTRATOR
Growth has been sluggish, but you could invest in the Hedgerow Fund. We'll get Gardener to trim it a bit. In the future, we will be planting cities.

The baby starts to cry and GARDENER sings a song, a rap, whilst digging.

GARDENER *(rapping softly)*
I speak truth to power
All power to the truth

And come the hour
You know I got proof

They may wanna twist it
To suit their delusions
They may wanna hide it
Disguise it thru' confusion

But me
I speak truth to power
All power to the truth
I dig for data
The only way ta
Get to treasure
Is hollow this crater

Yeah, I'm a whistleblower
I'll blow thru' the roof
And when I speak
Its straight up critique
Like Snowden I'll show them
Out in the open
Whose interests are favoured
When the facts are flavoured?
Transparency...
Is the enemy
Feared by...
Autocracy.

But me
I speak truth to power
All power to the truth
And come the hour

Disk Disposal

> You know I got proof
>
> You call yourself ethical
> I know you're very flexible
> Lets see you be credible
> Your actions are the recipe
> Could make your words edible
> Your claims to democracy
> May lead to hypocrisy
> Public discourse
> The only recourse
> Power is a cheater
> The truth is sweeter
> Exposed and naked
> Like a footprint in the snow

ARTIST *(drawing)*
The proportions are really something. And the dimensions. There's real depth.

ADMINISTRATOR
The potential is huge.

ARTIST *(signs his drawing of the hole)*.
I will trade you this signed artwork for the whole hole.

ADMINISTRATOR
What! and sell off this child's future debt? I don't think so.

ARTIST
Ok I'll just dig my own hole.

ADMINISTRATOR

Well we could all dig our own holes, sure, but then you'll devalue everyone's hole, including your own.

ARTIST

Tell me more about the Hedgerow Fund...

ADMINISTRATOR

It's a diverse portfolio, 60 percent hawthorn, blackthorn, and hazel, reducing exposure to riskier shrubs. Come I'll show you.

They leave Gardener as she continues to dig.

INT Sometime later ARTIST, ADMINISTRATOR, COLLECTOR are standing inside a large cavernous hole. Its an exhibition opening for ARTIST. They are looking at a framed drawing of the hole. Others are mingling.

COLLECTOR

And the picture? What's its value now?

ARTIST

I was able to swap it for a share of the hedgerow fund which I then in turn swapped for the hole.

COLLECTOR

Well, in spite of everything you were able to get on in life!

ARTIST *(modestly)*:

Oh not very far, you know, but nevertheless, its better than nothing.

COLLECTOR

Better than nothing! Is that even possible?

Disk Disposal

SCREEN

      SYSOP *(types)*
We have some holes in the server. Most likely old PHP scripts. Tell me if you need any of these otherwise they will get disabled or deleted. Hope all is well.

Ref: IR0020, IR0021, IR0022, IR0023
Name: DISK DESTRUCT
type: PHOTOS 3x, AUDIO TRANSCRIPT
Date: December 2013

The photos reveal failed attempt to open the disk and its subsequent destruction which looks almost ritualistic.

There seems to have been contradictory impulses around opening the disk up and keeping it unaccessible.

— What was no moved about the data in there?
— Did the disk burial take place in the jungle?

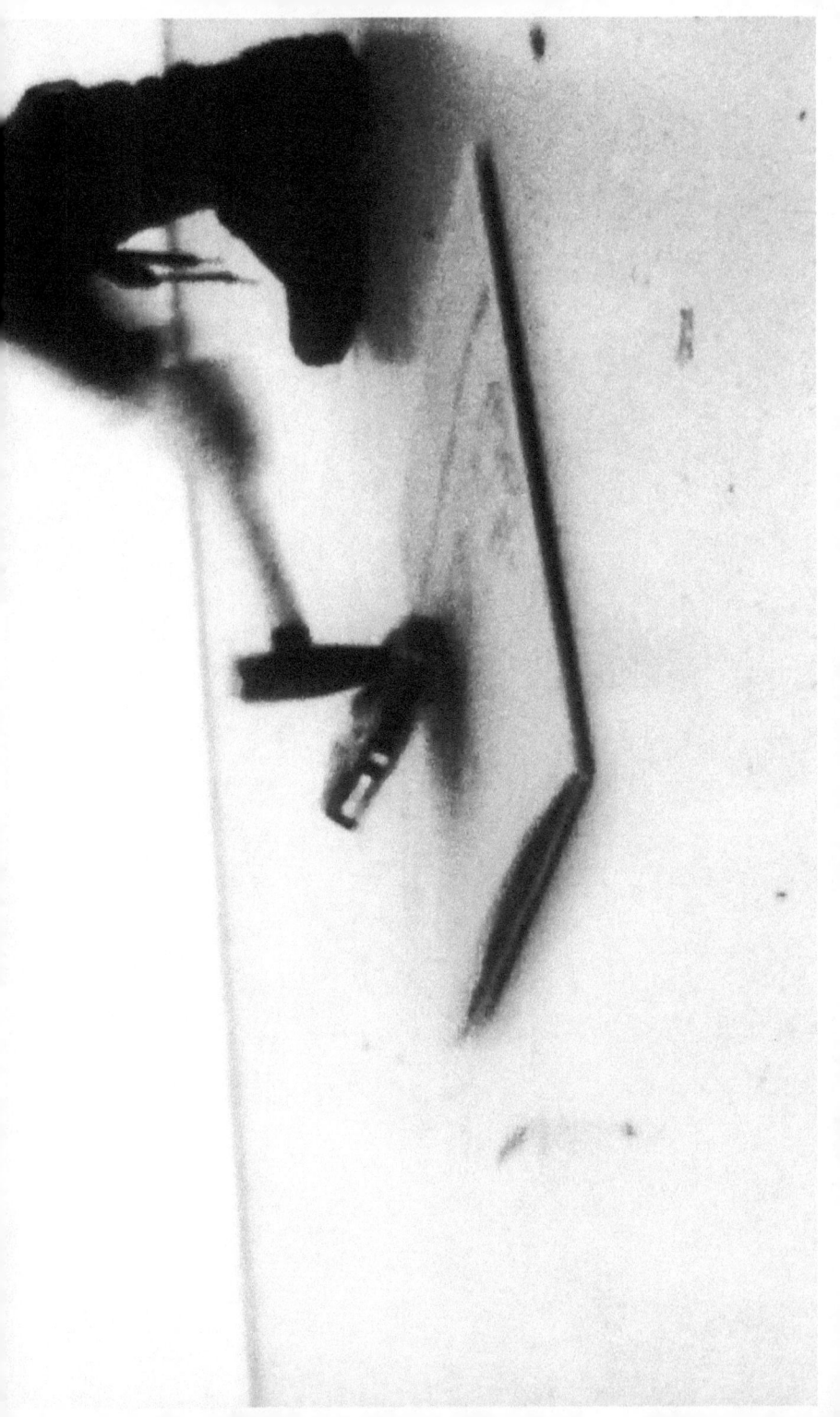

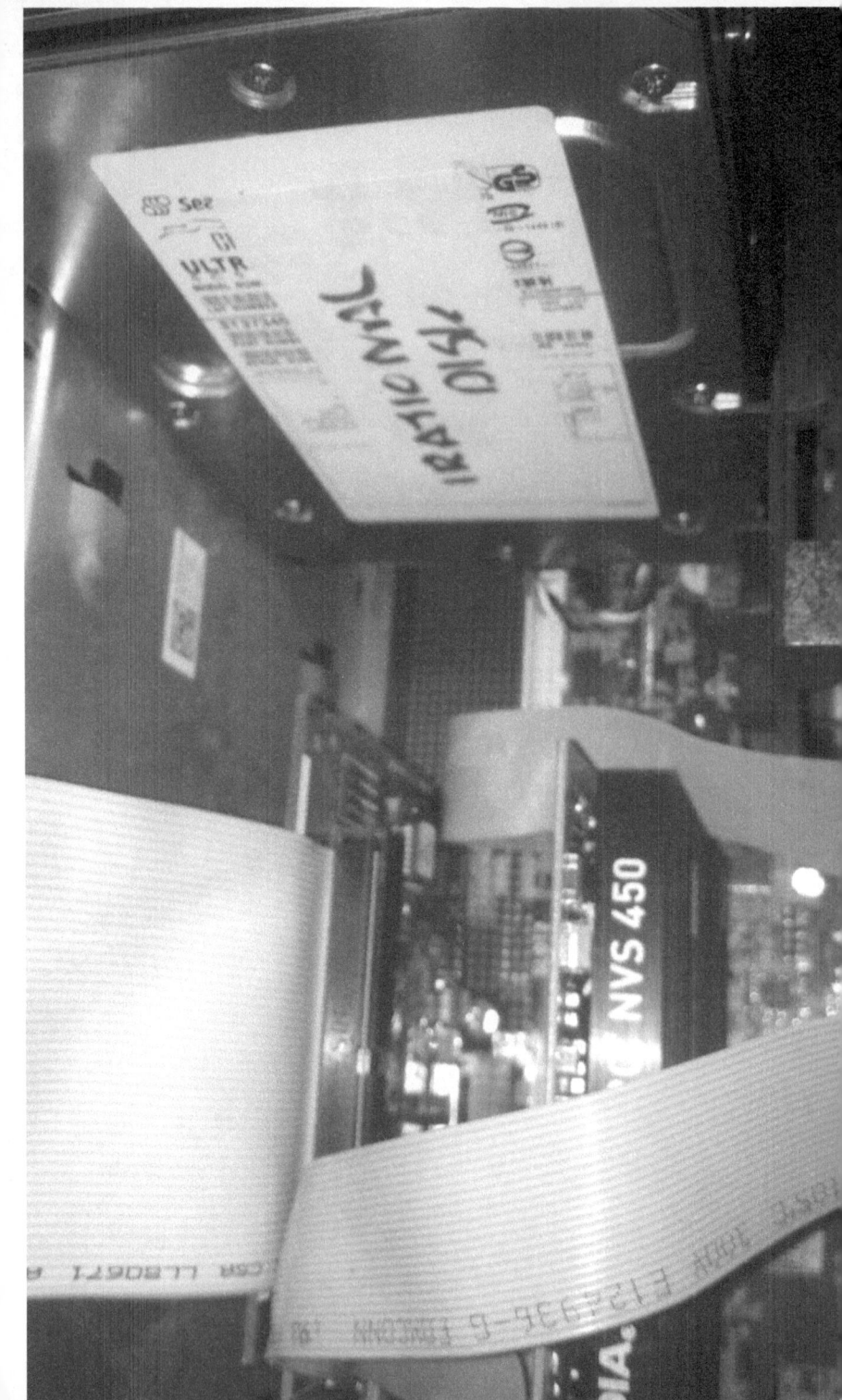

R.IR0023 - Not working

Treasurer: it does not work with this newer computer
Unemployed park attendant: oh, so it does not work
Treasurer: it's the power. this is one type of scsi, this is another, so it translates from there, so we would not be able to plug this disk in to this computer at all, we need this scsi connector and scsi pci bus, yeah. it is quite telling that, you know, without power none of this works (laughs)
Unemployed park attendant: yeah *(laughs)*
(rest of people look worried)
Unemployed park attendant: so there is no electricity for the disk
Treasurer: no, no, we need an older computer to plug it into..
Volontary: yeah, usually you need new stuff.

Ref: IR 0016
Name: RESIGNATION
Doc type: email

The Treasurer resignes in protest. Perhaps unresolvable disputes about distribution of funds. Indicative of administration malaise.

A common phenomenon amongst ant-mover groups where the presence of funds can be more damaging than an absence.

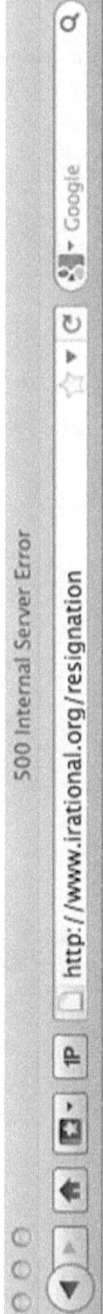

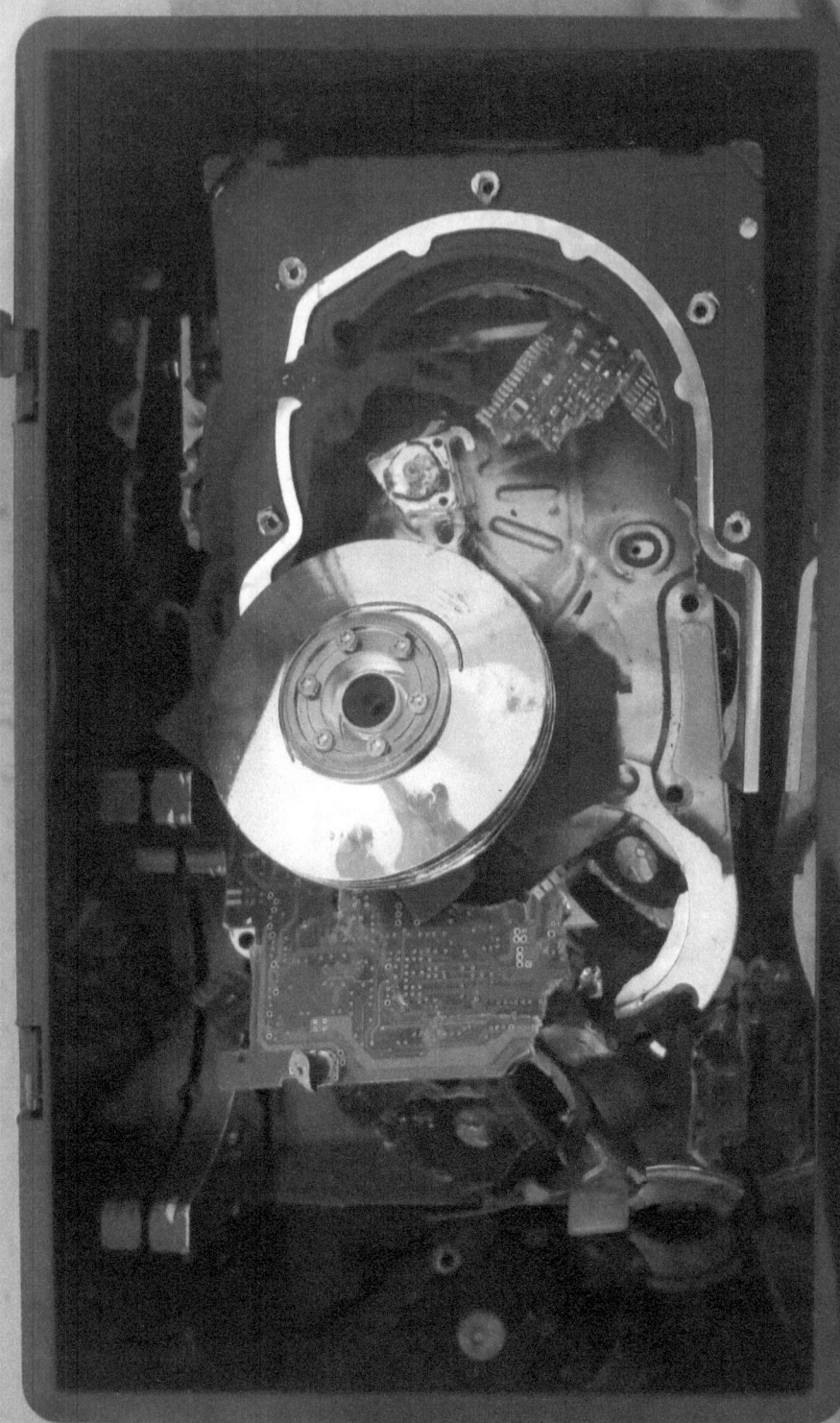

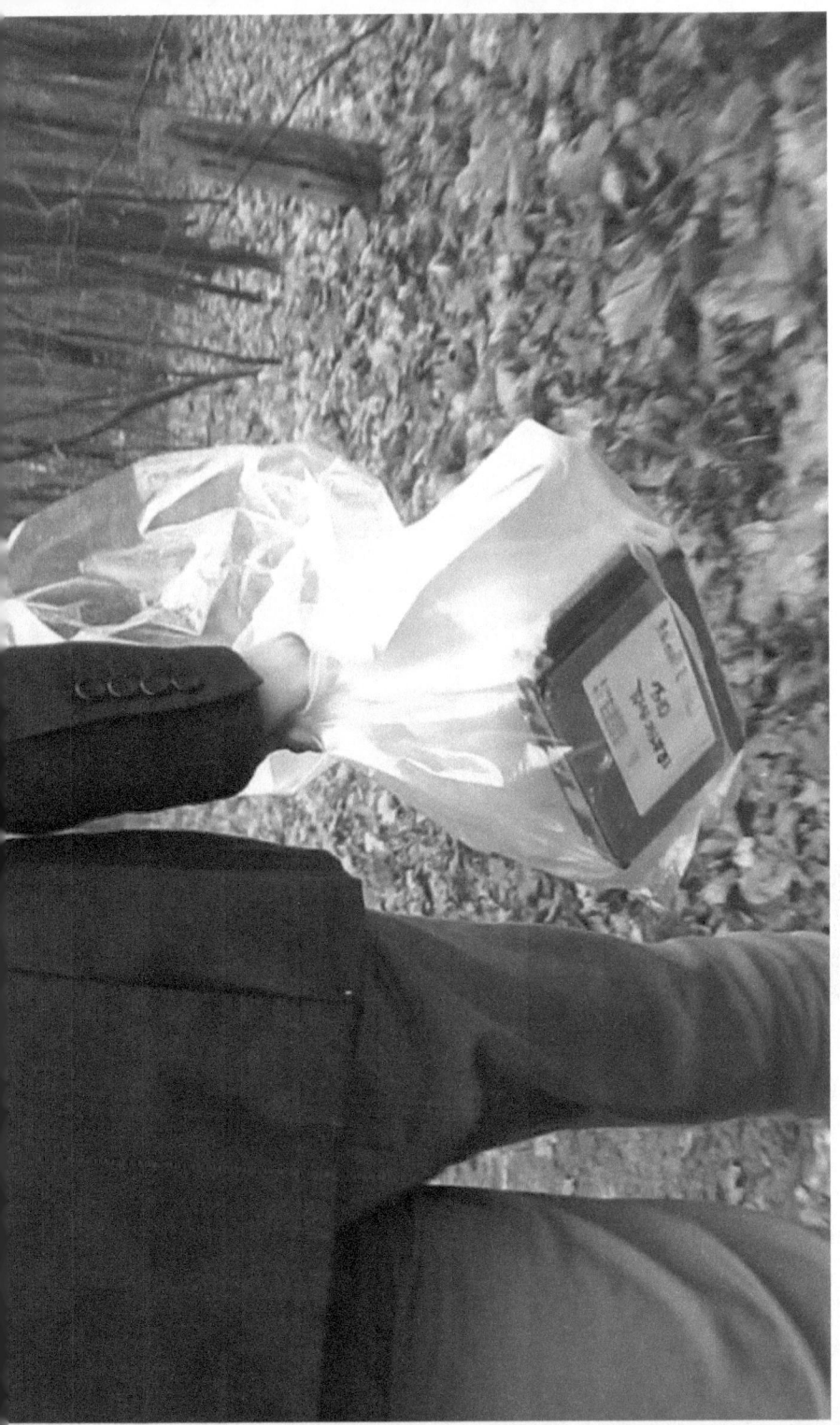

Ref.: IR 0024
Name: Apocalyptic scenario interviews
Type: transcript from audio interviews from my own archive
Date: 7th of December 2013

This conversation is a transcript of a recording I made in December 2013 which I had an opportunity to meet and spend some time with the members of Irrational.org during their visit to Limburg. As years passed I was struck with the "truth" contained in Treasure's words.

R.IR0024 - Apocalyptic scenario, interview (transcript).

V.M.: ...and, would there be someone who could use computers?
Treasurer: oh, no, no, for sure, there won't be computers around for sure in a few years
V.M.: yeah, so its all irrelevant what we are doing here
Treasurer: look really its relevant, we are looking what we can do with this downgraded technology, can we revive it, or can we just take it to the forest, smash it up and turn it to blades, or effective devices
V.M.: but why would worms need blades?
Treasurer: but we might be around in 15 years, you see..
V.M.: ah, for the last 3 years we might need blades
Treasurer: cannibalizing machines, cannibalizing humans
V.M.: what does it matter 3 years less or more of living for someone?
Treasurer: well, would you want to die now? i think anybody would do anything for one second longer unless you are in intense pain
V.M.: its like a competition of who will live longer
Treasurer: well i think if you are gonna observe, if humanity is gonna observe its own extinction. i am talking about not just a death of ourselves, and our loved ones, but the death of all species...
V.M.: you think its a good feeling to see the death of your loved ones around you?
Treasurer: no, no, no, but if we are gonna experience the death of humanity, than we need people who are going to observe that
V.M.: aah, to testify
Treasurer:... who are not the mafia. i would say as artists we have a sense of self preservation and also duty to humanity, and i think we should not give the planet, even if its just for few extra years, to the mafia. it should be given to the indigenous people, people embodying indigenous values or history. so if humanity has been around for last 500,000 years there is a great wealth of knowledge, wisdom

and tradition there which isn't carried by the mafia. so they shouldn't be the final representation of our dying species. thats why i train artists survival techniques

V.M.: to be the last survivors?

Treasurer: yeah, generally the people who.., generally are on the margins, anyway, most precarious, and generally the first people that have to leave. and when super-states fall apart, the states or town or families, its generally the artists that have to go first, to run away. so when society.. well its already collapsing, you see, i have a personal crisis now... then other people in my position have financial crisis, political crisis.. personally i am facing potentially moving in to the forest. your generation from your part of the world did have to run away and live in the forest, or they had to cross the borders to go to other countries...

V.M.: but do you think people should become selfish in those situations?

Treasurer: no, no, no at all.. but what i am trying to say is - this crisis is being caused by the mafia, and certainly should not cede the humanity to them in the last moment.

V.M.: do you think those artists could try to preserve humanity?

Treasurer: i don't think there is any chance to preserve humanity. even fukushima, they say there are 4 nuclear stations there. there are, what, 400 nuclear power stations in the world? even japan is not able to solve 4 nuclear accidents, they don't have the technology, they don't have the economy. fukushima has almost doomed the northern hemisphere within a next few years to extinction. think of all the other power stations that there are. there is not enough energy, there is not enough resources, not enough knowledge to decomission them and cope with the nuclear waste, and nuclear waste lasts at least another 100,000 years. so even just nuclear power itself is the death of humanity, and many species. and then you have also climate change which is dooming 200 species a day to extinction. 200 species a day go extinct because of climate change. and you have peak oil which is going to cause the extinction of western civilisation as we know it. people won't be able to afford to do any of this stuff. you'll be back onto feudal agriculture if you are lucky, if you can go back to the ground that has been destroyed by industrial agriculture. most of the fields in my country cannot exist without fertilizers. or you can describe agriculture now as a technology for converting oil into food. so without oil we dont have agriculture. so peak oil sentences humanity to extinction or the huge section of it. financial crisis is just an effect of that. and climate change threatens to extinct every human and multiple thousands of species, and nuclear power also, tens of thousands. all of these things are coming together and impacting civilians, in the next 5 to 10 years..

V.M.: so can you give some message to the young artists?

Treasurer: young artists? so, the economy is never going to get better, technology is not going to get better, its only going to get worse, environment is going to get worse, civilisation is going to collapse..

V.M.: short message

Treasurer: short message to young artists? make sure you get paid!

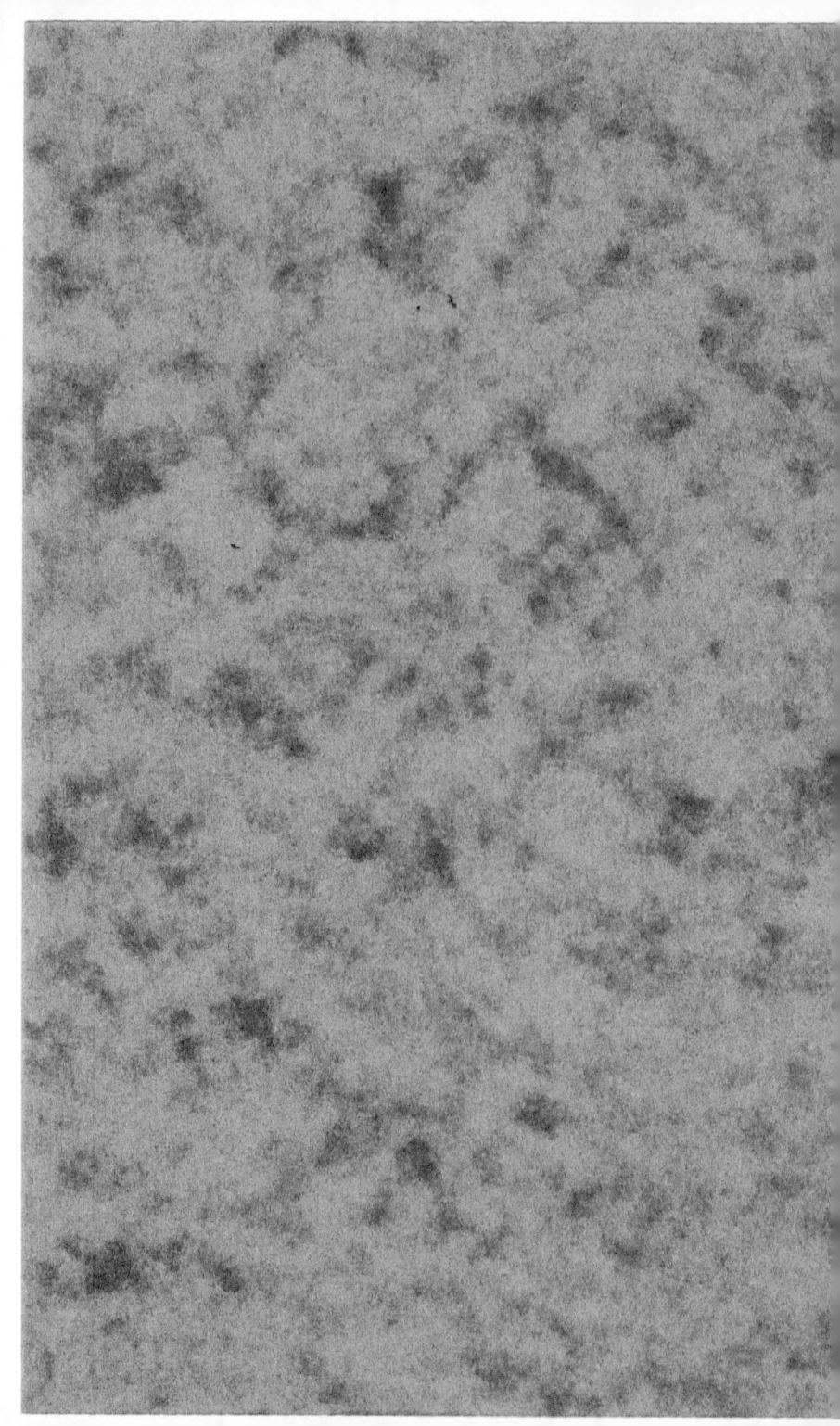

FOREST BUREAU

Ref: IR0017
Name: IRATIONAL.ORG COMMON OBJECTS
Type: Photocopies of objects and their descriptions
Date: December 2013

This collection and their analytical descriptions demonstrate attempts to translate stark data into materiality; access subject/object relations towards the dark.

IR0017.01/02 2× USB DRIVES / Memory storage devices. On each, the fails operating system, used to disguise velocity on internet. Indicates interest in privacy and in sharing. Only the Syssop and the Treasurer had access to root of Irational.org.

IR0017.03 OYSTER CARD With RFID tracking technology, used for travelling the London subway system. Indicates the mechanism of accuracy and accessibility in which the dark/programmation protocols were set.

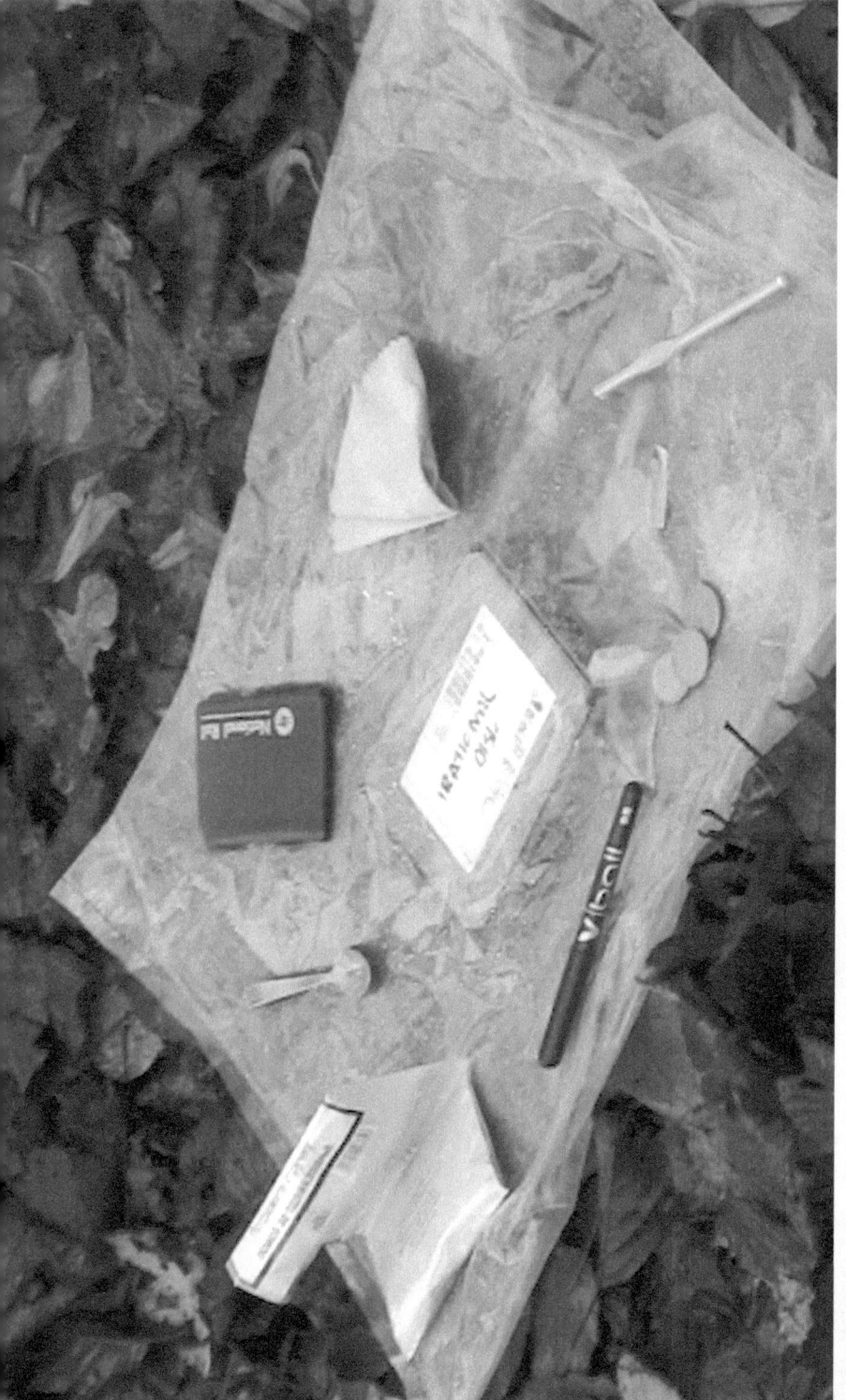

IR.0017.04 SPOON Assumed to have belonged to Gardener. Emphasizes never as source of addictive income, also dependency on external institutional funding.

IR.0017.05 CROCHET HOOK. Made of aluminium. Probably belonged to Thief. Indicates artisanal aspects of the sink/organization, implies manual wearing of data, and materials, and relations.

IR.0017.06 TOBACO BAG. Cleaning lady was the only straw user in the group. Indicates addictions and dependencies around work, the server, the dark, & the group.

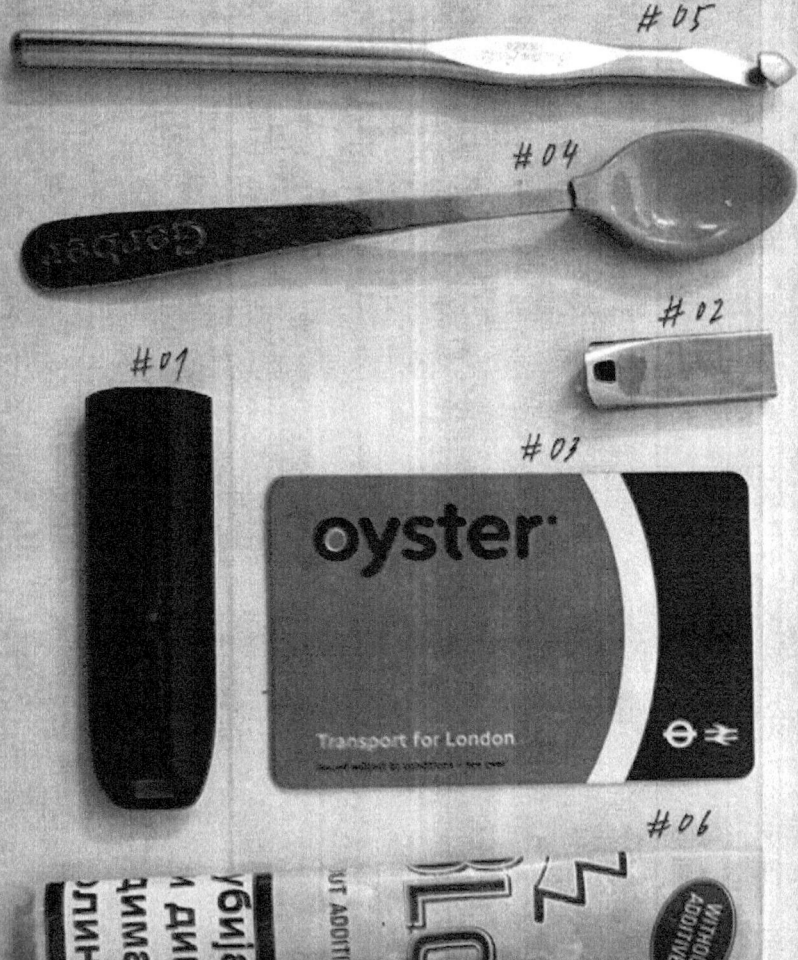

OBJECT ANONYMOUS OPERATING SYSTEM & PROGRAMS

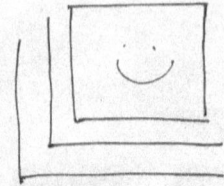

DESCRIPTION:
PEER REVIEWED OPEN SOURCE SOFTWARE TO DISGUISE IDENTITY OF USER

RELATION:
USER

UTILITY:
PRIVACY / ANONIMITY

OBJECT **USB STICK**

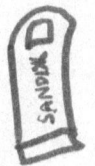

DESCRIPTION:
METAL SURFACED
DISTINGUISHED FROM OTHER
SIMILAR USB STICKS BY NAIL
POLISH DOT.

RELATION: BOUGHT OWNED, PRACTICAL
UTILITY. LIGHTWEIGHT
ALSO DISPOSABLE/INTERCHANGEABLE.
EASY TO SECURE TO KEY RING, NOT LOSE

UTILITY:
COMPUTER OPERATING SYSTEM
STORAGE
+ MEMORY

OBJECT _Oyster Card for London trav[el]_

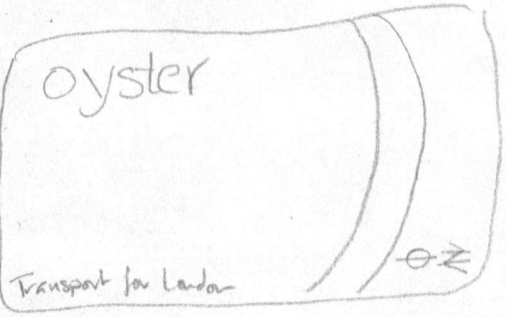

DESCRIPTION:
A card, plastic, blue
unregistered, unique RFID

RELATION:
Access to travel, in London by bus + train value-holding, a tracking device, a hostile object, part of larger control mechanism, debt. on the travel network

UTILITY:
Access to travel

OBJECT _BABY SPOON._

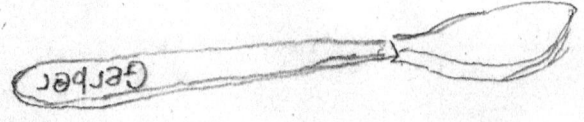

DESCRIPTION: a spoon like a tea-spoon, but for feeding babys with. metal handle with Gerber etched into it. the spoon end is made of a soft rubber, which is pink in colour.

RELATION: Given to us by Daniels Cousion, I dont leave the house without it. I like it but would not be devasted if I lost it.

UTILITY: to aid the safe and easy feeding of a baby who has no teeth or skill in spooning up their own food and putting it in the mouth

OBJECT stolen crochet needle

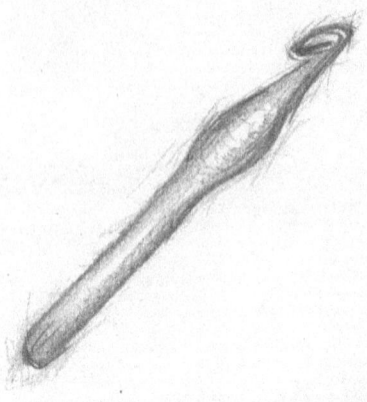

DESCRIPTION: metal, light-metalic green, elongated.
has a hole on one end
gets wider and flatter a little bit above the middle

RELATION: always on me, related to something I am very skilled at and enjoy doing at any given moment. stolen at a huge craft store in Maryland, close to my parents house (chainstore, multinational)

UTILITY: used to turn anything stringlike into fabric, textile. one of the simplest weaving devices. meditative, used to demonstrate the relation between repetition and variation.

OBJECT TOBACO BAG

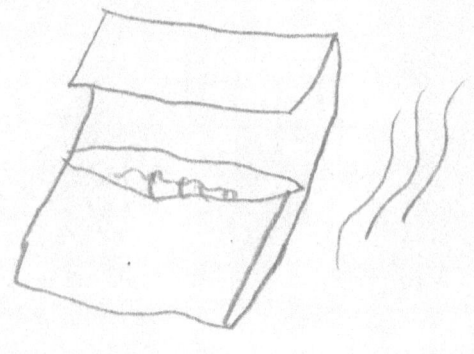

DESCRIPTION:
TOBACO LEAFS, GROWED DRIED, FERMENTED, PACKAGED NO ADDITIVES, BOUGHT IN THE KIOSK

RELATION: DISSPOSABLE BAG, TOBACCO CONSSUMED, CONSUMPTION

UTILITY: PLEISURE, ADDICTION

CEO
OF THE
COMMONS

EXT: *The Lüneburg Heide.*
SET: *A juniper bush, a fire, an embroidered tablecloth.*
From an aluminium security briefcase CEO pulls out 8 x objects and places them around, but not on, the tablecloth -
2 x snails with white marks on their shells
a crochet hook
a spoon
an Oyster travelcard
2 x USB sticks
a pouch of tobacco

CEO and SECRETARY sit together around the tablecloth.

(Stage directions: use live snails where possible, and if they try to escape then let them, especially if they traverse the tablecloth)

CEO

Hello dear friends, comrades and enemies. Glad you could make it. As the Chair of this meeting I would like to welcome you all to the Commons Committee AGM. I think we should register who is here. Secretary, will you take the minutes?

SECRETARY

Don't worry, I'm committing everything to memory, minute by minute.

CEO

Good. Make sure you have a back-up. So whose present today? Editor, Gardener, Host, Therapist, Lawyer, Artist. Have there been any apologies, Secretary?

SECRETARY

Everyone is very sorry.

CEO

How about Declarations of Interest? Any interesting declarations? Anybody? I have just one - I declare an end to representation!

SECRETARY

Isn't it 'Conflicts of Interest' that must be declared?

CEO

Conflicts are always of interest. Are there any conflicts? Nature vs Capital. Is it on the Agenda? Where is the Agenda?

SECRETARY

We had one but you used it to light the fire. I suggest we make it up as we go along.

CEO

So Capitals vs Nature... where are we with that?

SECRETARY

Capital insists that the Commons is in scarce supply and Nature insists the opposite, it is abundant. The tree-climbing report might be pertinent to this.

CEO

Ah, yes the tree-climbing report submitted by GARDENER. Lets see.

SECRETARY hands CEO a piece of paper

CEO
'A silver birch, 35 ft up, wrapped myself around the new branch, grabbed wildly at the twigs, let myself hang down, banged the edges of my shoes against the bark, searching for stairs, I took a razor sore from the bark on my descent, searched for the ground, and stood.'

SECRETARY *(clapping).*
She sustained an injury.

CEO
How much is this worth? Have the cost-analyses been done on this abstract labour? I think we can include this in the Annual Report and prove Nature's point – we can be culturally abundant with natural resources.

SECRETARY
This will add value to our Commons Culture Index.

CEO
I have some proposals I'd like to share. In the past when proposals have been put for vote, silence is taken to mean consent. I see no reason to deviate from this. Anyone object?

CEO looks around at the dumb objects and the snails.
A horsefly buzzes around CEO's face. SECRETARY slaps CEO's face and the buzzing stops.

CEO
So, I propose we trade in 400 juniper berries from the bank for 400 oz of salt. In this time of conflict we need to diversify. And salt's value has soared, passing 1,000 junipier berries an oz on Wednesday afternoon.

SECRETARY

Really? It was only 400 an oz on Tuesday afternoon.

CEO

The saltmines on the Commons have been working round the clock to keep up with demand.

SECRETARY

Soon we won't have any juniper berries – just salt.

CEO

We need an alternative to juniper berry hegemony. And salt is open source. And finite. And requires more labour.

CEO takes a handful of Juniper berries and fills his mouth, munching vigourously, leaving berry juice stains all over his face

SECRETARY

So what happens if the miners give up mining?

CEO

If half the miners give up, mining becomes twice as profitable, and they are therefore incentivised to keep with it. In other words, the number of miners will reach an equilibrium such that it is barely profitable.

SECRETARY

Well, as long as you feel confident. Its all about confidence.

CEO

Lets put this to the vote. All those in favour say 'Eye'

SECRETARY

Nose

CEO

Ah, an objection?

SECRETARY

Sorry wrong anatomical affirmation. Eye!

CEO

Wonderful, consensus has been reached.

And remember if you objects object, you are free to leave at anytime, although you cannot take any berries or salt with you. Any other business?

SECRETARY

I have a postcard that arrived yesterday from Sysop, he says

'On the beach, among the planks, suburban driftwood, brought by storms, while a whole skyful of starfish slowly swarm about me. We are building our dragon ships with no roof for the timid; the sail is already billowing on the first mast. Two of the boats will set out first as a scouting party, as our raven, our knights of grail, as leader of the men, of the mission, I will follow a few days layer, with the six remaining ships. My next epistle has already been sent to all of you my friends...'

...then the text becomes confused and unreadable, it was sent two years ago...maybe he drowned?

CEO

I excommunicated him, remember?

Forest Bureau

SECRETARY

Are you sure he didn't drown. There have been many drownings recently in the marshes. Social outcasts.

CEO

We use the landscape to punish the troublemakers. Move them out to the marshes. Is it raining or are you crying all over me?

It starts raining. CEO and SECRETARY gather all the board members and put them back in the briefcase. The fire is extinguished.

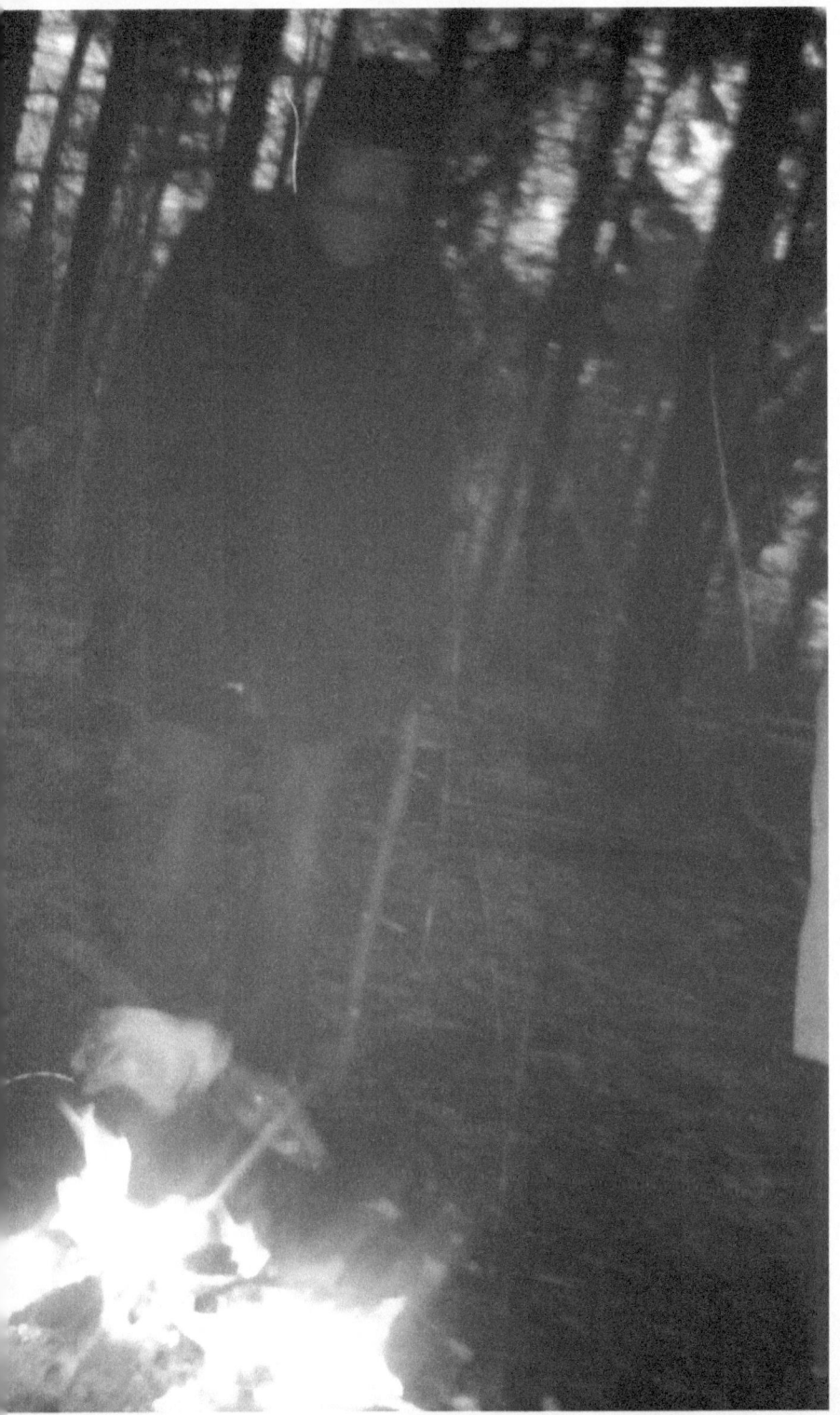

Ref   the word hung from the highest tree
      the forest hides from the tree from the highest
        branch at the head of the word

      the branch sticks out its tongue at the heart
              of the word tree

      the head branch runs to the head the tree
        chopped off lost head errant heart
       drunk with word the forest hides the tree sticks
                          out its original tongue

      no head without shadow chopped off
                 lost head errant heart

       the word earth planted in the word
              the word edge at the heart

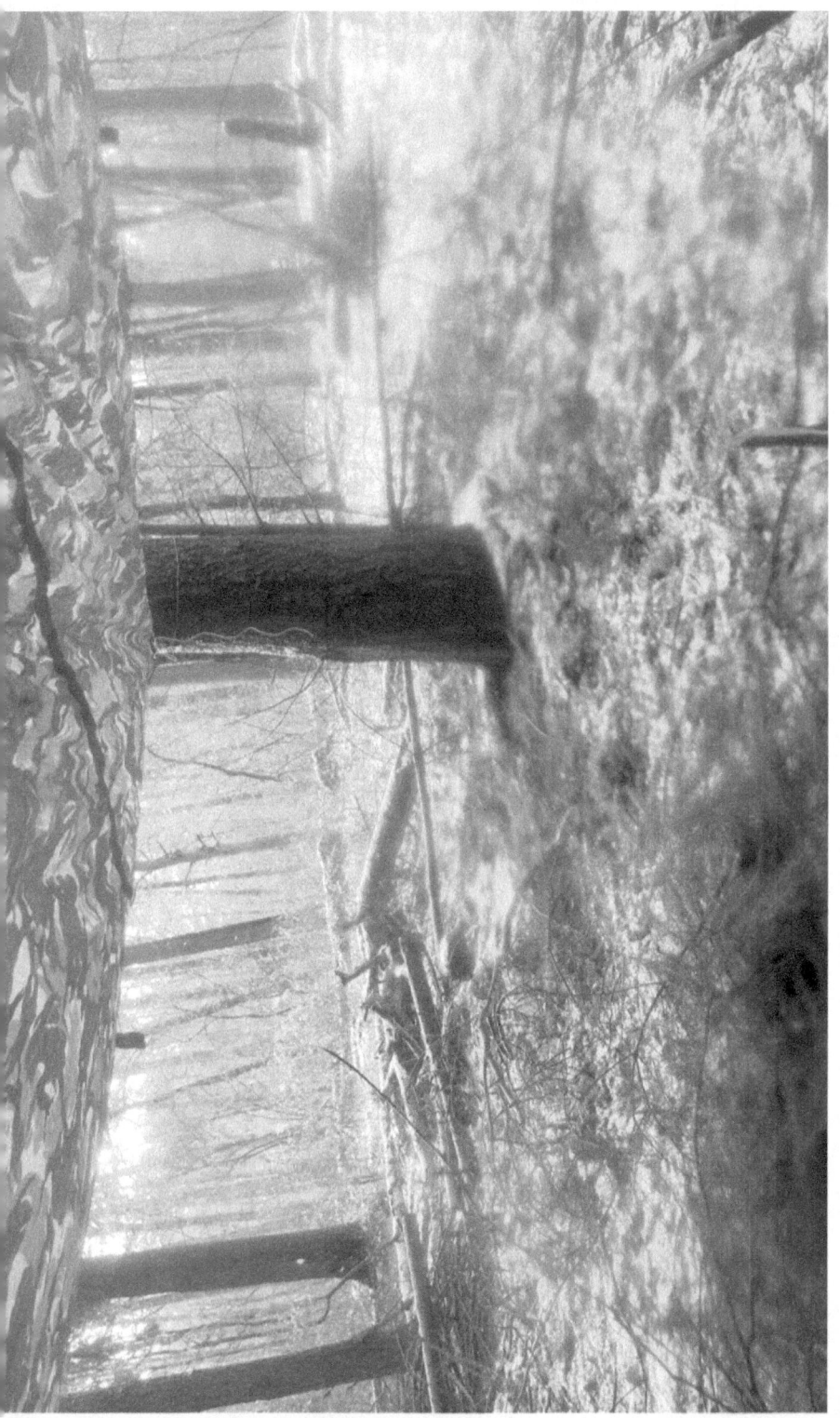

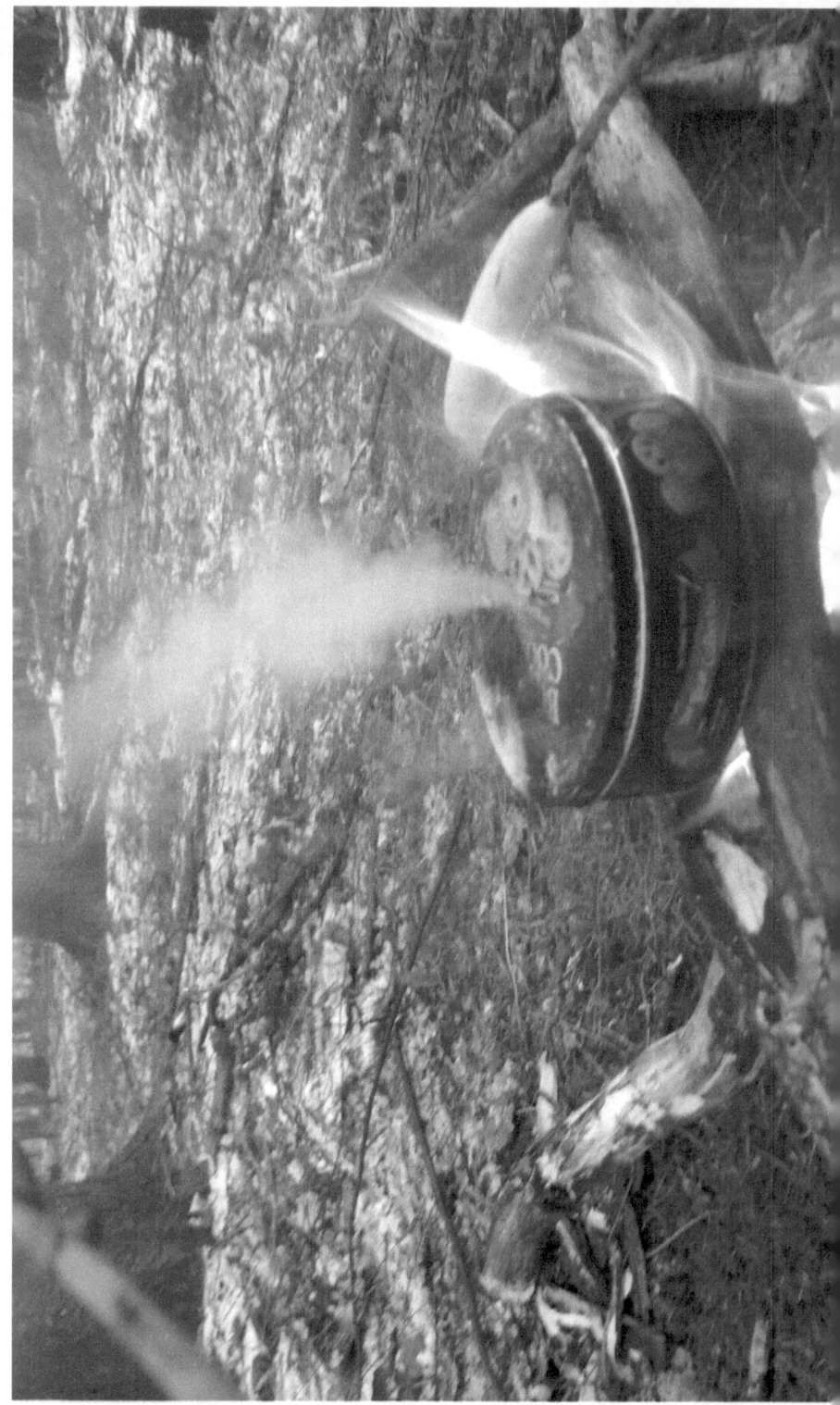

R.IR0027

Thief: this is a document?
Treasurer: yeah, that i am already burning
Thief: burn the documents before we put secrets on them?
Treasurer: yeah..
.....
Thief: how do you make charcoal, in the can?
Treasurer: yeah, so you need to heat it up, but not burn it, the charcoal needs to be basically baked, the wood needs to be baked
Treasurer: so what you do, is put it in the tin, make a hole. things like this are very useful in the forest. stop animals eating the stuff, all sorts of other things..
....
Treasurer: whose gloves are those?
Thief: mine.
Treasurer: ok
Thief: you think they'll burn?
Treasurer: i see many people burn their gloves and socks in situations like this..
....
Treasurer: just watch it, and when it goes out
Thief: is it different looking? does it change?
Treasurer: yeah it does, but you can try and light that with a small burning stick, to see if it ignites
Thief: and if it ignites, then it is wood gas
Treasurer: yeah
Thief: and then it's charcoal?
Treasurer: yeah, but you have to wait till it stops, let's see
Thief: what kind of gas?
Treasurer: wood tarts. it's not lighting up, i don't think

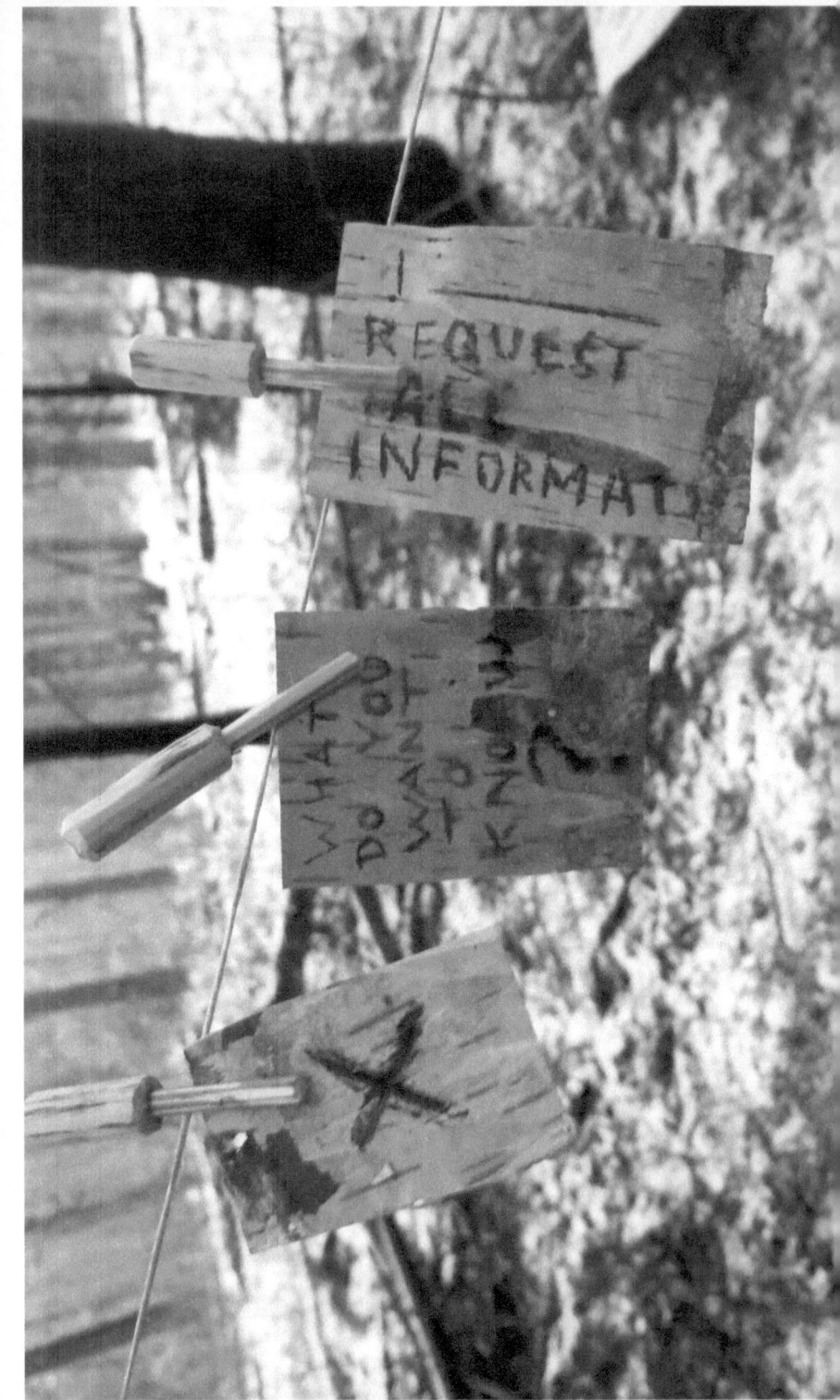

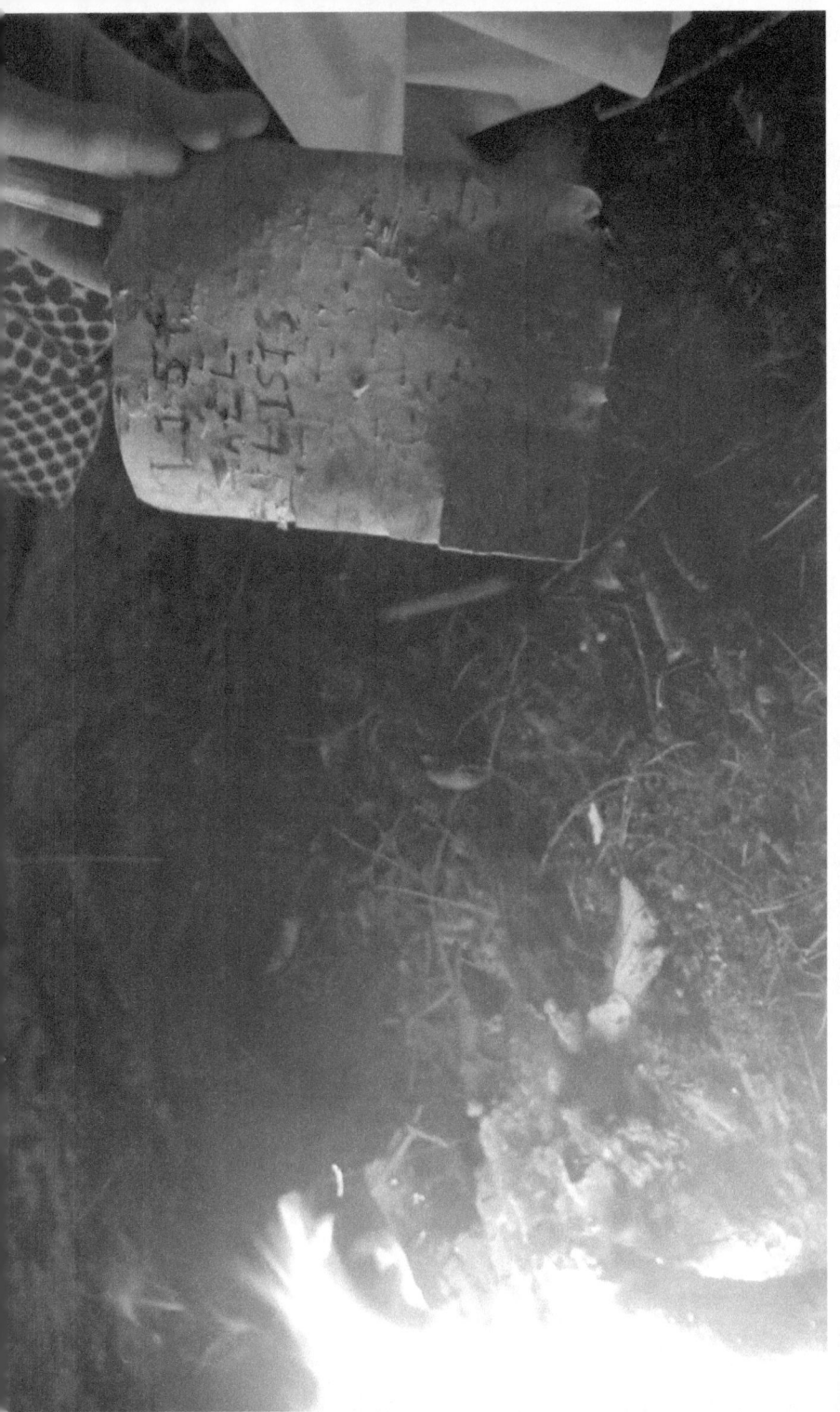

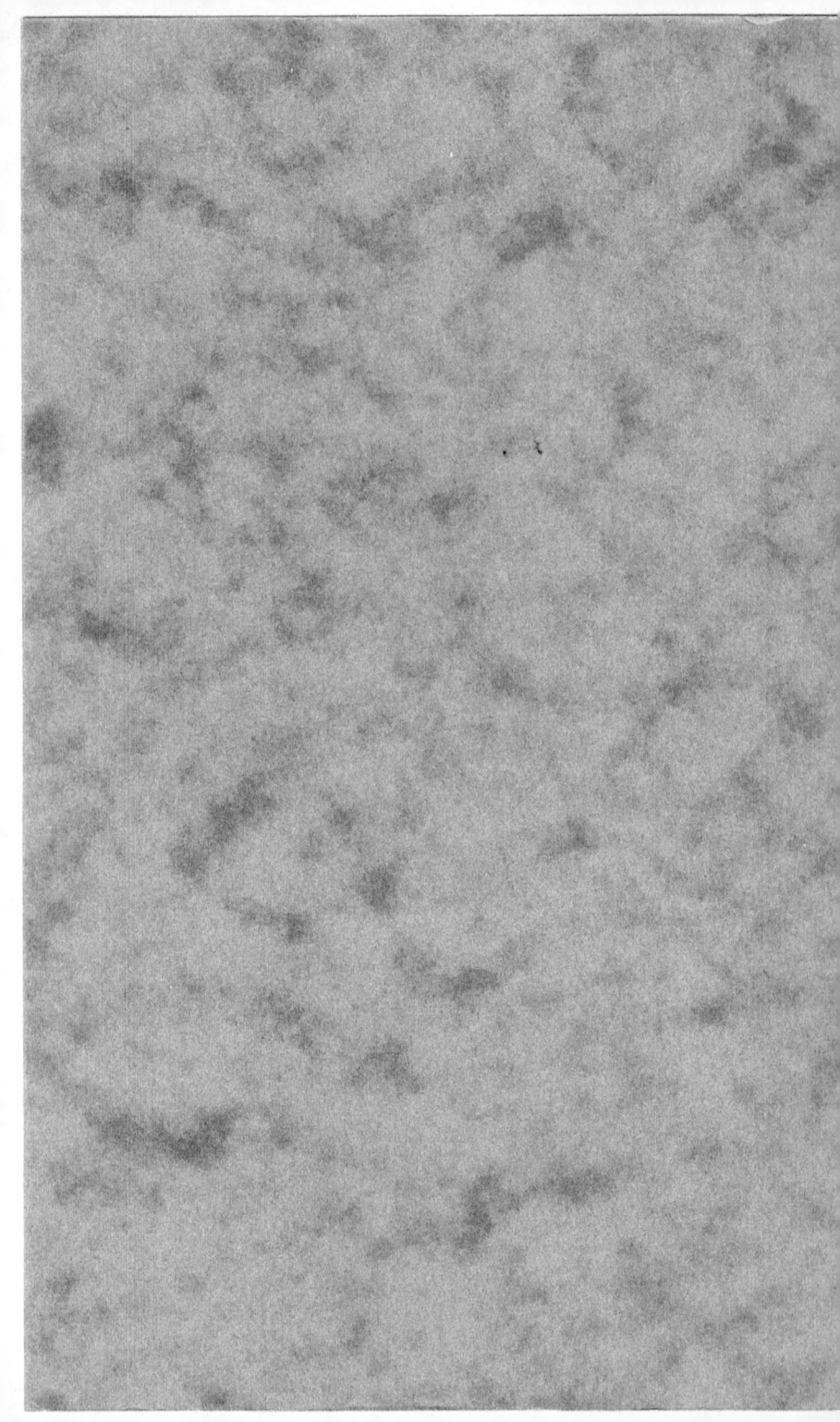

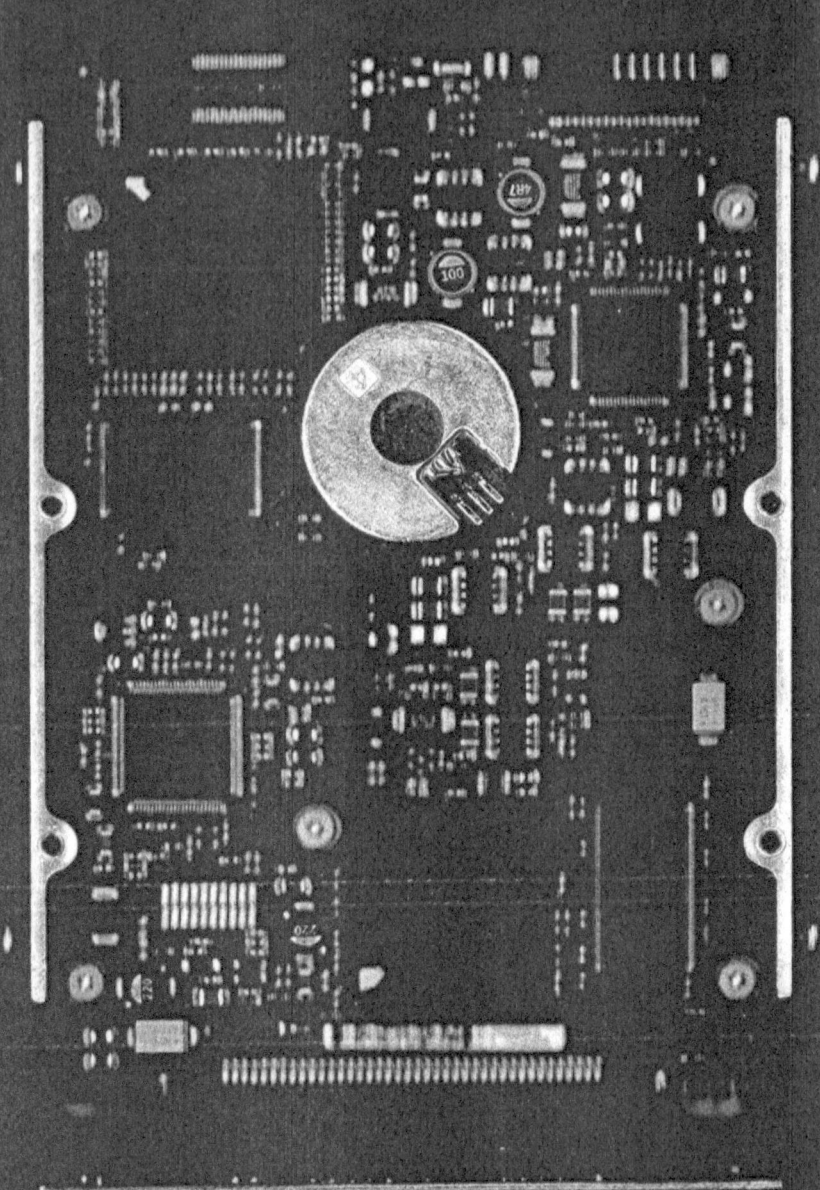

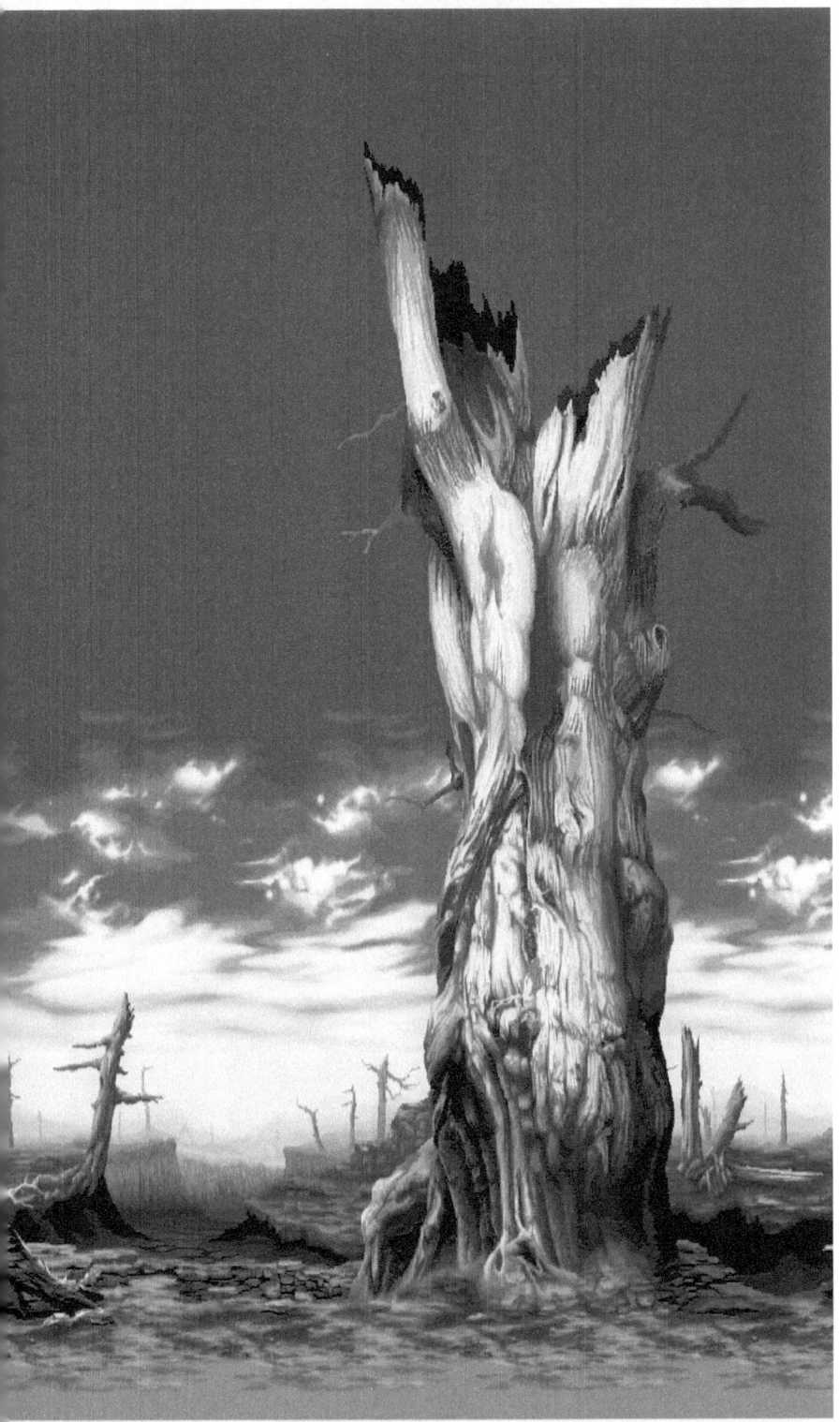

www.ingramcontent.com/pod-product-compliance
Lightning Source LLC
Chambersburg PA
CBHW020436220526
45464CB00002B/729